THE HEIGHT OF PROPHET ADAM

At the Crossroads of Science & Scripture

بِسْمِ ٱللَّهِ ٱلرَّحْمَٰنِ ٱلرَّحِيمِ

THE HEIGHT OF PROPHET ADAM

At the Crossroads of Science & Scripture

Muntasir Zaman

with an afterword by Jonathan A.C. Brown

BEACON BOOKS

First published in the UK by Beacon Books and Media Ltd
Earl Business Centre, Dowry Street, Oldham OL8 2PF UK.

First edition published in 2022

www.beaconbooks.net

ISBN 978-1-915025-32-6 Paperback
ISBN 978-1-915025-33-3 Hardback
ISBN 978-1-915025-34-0 Ebook

Cataloging-in-Publication record for this book is available from the British Library

Cover design by Zaki Ansari and Lily Martinez

In loving memory of the ʿ*ulamā*ʾ who lost their lives to the Covid pandemic —

"*Rather, they are alive, but you perceive it not*" (Q. 2:154)

"Even if [the hadith] were sound, it would be easier to interpret it metaphorically than to reject matters [of astronomy] that are conclusively true...The greatest thing in which the atheists rejoice is for the defender of religion to declare that these [demonstrable facts] are contrary to the faith. Thus, the atheist's path for refuting the religion becomes easy if [accepting] such views is required."[1]

Abū Ḥāmid al-Ghazālī (d. 505 AH)

1 Translation adapted from al-Ghazālī, *The Incoherence of the Philosophers: A Parallel English-Arabic Text* (Provo, UT: Brigham Young University Press, 2000), 7.

CONTENTS

LIST OF FIGURES

NOTE ON CONVENTIONS

The word "hadith" will not be transliterated, and it will be used for both singular and plural forms (e.g., this hadith, these hadith). The Arabic definite article *al-* is not added to non-Arabic affiliations of scholars who lived in recent centuries (e.g., Dihlawī). Terminal *h* representing the *tā' marbūṭa* in pausal form for Arabic words is generally omitted (e.g., Taymiyya, rather than Taymiyyah), but final *hā'* is retained in the transliteration of words in non-Arabic contexts (e.g., Khānah). Dates after 1750 are generally given according to the Common Era. For dates before 1750, only the Hijrī date is given.

ACKNOWLEDGEMENTS

None of this research would have been possible without the advice and guidance of experts from a wide array of disciplines: scientists, archaeologists, historians, Biblical scholars, hadith scholars, linguists, and many others, all of whom helped me to better understand their respective fields. I am indebted to Shoaib Ahmed Malik, Nazir Khan, Saif ul Hadi, Haroon Anis, and my students at Qalam Institute for their valuable feedback on the draft of this work. Sharif El-Tobgui's attention to detail and exceptional editorial skills have improved the quality of this book immensely. Zaki Ansari went out of his way to help design the cover and the diagrams. Muhammad Hozien never hesitated to answer my pedantic questions.

My loving wife is a standard of care, affection, and dedication that I can only dream of reaching. I may have done the research and writing, but she has put in an equal, if not greater, number of hours in making this book a reality. Husna and Salman, my self-appointed editors, bring immense joy to my life and have put a smile on my face during the most trying moments. The prayers of my parents are what allow me to produce anything of value; no amount of words can express my gratitude for their willingness to support all my educational endeavors regardless of the sacrifices. I am unable to fulfill the debts that I owe to all those who have generously shared their time and advice, so I pray that Allah grant them a beautiful recompense.

PREFACE

Since the advent of Islam, Muslim scholars have ruminated over the place of reason in interpreting scripture and formulated methods for resolving any conflict arising between the two. The preeminent theologian and mystic Abū Ḥāmid al-Ghazālī provides a fitting analogy to describe the balance one ought to seek when interacting with these two integral epistemological sources. The light of revelation, he explains, is like the rays of the sun. A person who abandons his rational faculties and suffices himself with revelation alone is like someone exposed to the rays of the sun with his eyes closed; he is effectively no different from a blind person. Therefore, "the intellect coupled with revelation is light upon light. The onlooker with an eye blind to one of them is drawn in by a deceptive rope."[1] Though al-Ghazālī's analogy referenced the role of the intellect, it can be used to appreciate the role that other epistemological sources play in understanding scripture, such as empirical and scientific knowledge, which feature prominently in conversations about the viability and relevance of Islam in the modern age.

The conflict between scripture and science is a common source of anxiety for people of faith, and Muslims are no exception. The inability to justify certain matters of Islam in light of current scientific consensus has led some Muslims to doubt their faith. When confronted with hadith that clash with their firmly held assumptions about science, Muslims unequipped with the requisite hermeneutic tools feel the need to pass a judgment on either side of the debate. Some discuss scripture without any religious training, while others blindly express opinions about science. Yet, tension between hadith

1 Abū Ḥāmid al-Ghazālī, *al-Iqtiṣād fī al-iʿtiqād* (Cairo: Dār al-Baṣāʾir, 2009), 70.

and other disciplines is not without historical precedent. That many of the same hadith regarded as problematic today were already discussed in detail by the greatest Islamic minds is often overlooked in these discussions.

Premodern scholars were not oblivious to empirical realities nor blind to logical fallacies.[2] Many, in fact, were actively involved in the rational and scientific fields.[3] However, they had a more robust conception of scripture as a purveyor of knowledge, while their current-day detractors tend to assume an empiricist epistemology that diminishes the role of scripture in providing knowledge about the world. That said, traditional scholars understood that the strength and value of scriptural evidence varied. For instance, a solitary hadith with conflicting routes of transmission is not on par with an explicit verse of the Qur'ān; the Qur'ān takes precedence. Accordingly, they anticipated the need to resolve serious conflicts between scripture and empirical or rational proofs. A deep dive into one such case study can provide insights into the application of this traditional methodology in dealing with modern contentions. For our case study, we will examine hadith that (1) Prophet Adam was sixty cubits tall and that (2) mankind has since been decreasing in height. Both propositions are considered scientifically and archaeologically contentious.

This project began as a brief answer to a text message in 2018. It then became the focus of my research for the next few years, as I quickly realized that this was more than a straightforward hadith about the father of humankind. This hadith serves as an excellent

---

2 Hadith scholars often dismissed hadith with seemingly reliable chains if they contained a logical fallacy, empirically unsound information, or anachronisms, a process dubbed content criticism (*naqd al-matn*). See Jonathan A.C. Brown, "How We Know Early Ḥadīth Critics Did *Matn* Criticism and Why It's So Hard to Find," *Islamic Law and Society* 15 (2008): 154–162; Ṣalāḥ al-Dīn al-Idlibī, *Manhaj naqd al-matn 'inda 'ulamā' al-ḥadīth al-nabawī* (Amman: Dār al-Fatḥ, 2013), 316–339.

3 The Muslim scientist and physician Ibn al-Nafīs (d. 687 AH) is one example. Not only was he a skilled physician and scientist, second only to Ibn Sīnā (d. 427 AH), but he also wrote a book on hadith nomenclature. See Tāj al-Dīn al-Subkī, *Ṭabaqāt al-Shāfi'iyya al-kubrā* (Cairo: 'Īsā al-Bābī al-Ḥalabī, 1964), 8:305; M. Meyerhof and J. Schacht, "Ibn al-Nafīs," in *Encyclopaedia of Islam,* 2nd ed., 2012.

case study in dealing with contentious hadith because it weaves together heated topics that are at the center of modern hadith studies: science, archaeology, epistemology, the status of the two *Ṣaḥīḥ*s, renewed and competing methods of hadith critique, the *isrāʾīliyyāt*, and the transmission of the Companions, to name the most salient. Chapter 1 lays out the framework adopted throughout this study and explains key terms. Chapter 2 unpacks the scriptural, scientific, and archaeological arguments for and against the existence of human giants. Chapter 3 surveys classical interpretations of the hadith under discussion. Chapters 4–6 explore three important methods of dealing with the present contentions. The last chapter touches on the relationship between miracles and the laws of nature, while the conclusion summarizes the key findings of the study. Finally, Dr. Jonathan Brown has graciously contributed an afterword discussing the topic of the *isrāʾīliyyāt.*

Although the present research is geared towards the scriptural end of the conversation, its goal is to facilitate a nuanced dialogue on the relationship between science and scripture, in which hadith play a distinct role given their varied epistemic levels of transmission in contrast to the Qurʾān, the transmission of which is definitive. I have attempted to clarify some of the advanced terms and concepts employed in this book. However, the assumption in this study is that readers have some background in the Islamic sciences. To facilitate reading, subtitles have been added throughout each chapter even though certain sections are relatively short.

Muntasir Zaman
October 2021| Dallas, Texas

PART I

SETTING THE STAGE

INTRODUCTION

This book explores the interplay between science and hadith by examining reports on the height of Prophet Adam and subsequent generations, with particular focus on scholarly endeavors to treat the ostensible conflict between such reports and the extant scientific and archaeological data. I examine the hadith recorded by al-Bukhārī (d. 256 AH) and Muslim (d. 261 AH) that (1) describes the height of Adam as sixty cubits—that is, about twenty-eight meters, or ninety feet—and (2) indicates that human height has since been decreasing. At the outset, it may help to introduce the hadith studied in this monograph. There are multiple routes of this hadith, but one will suffice for now. Hammām b. Munabbih narrates from Abū Hurayra that the Prophet (ṣ) said:

> Allah created Adam with *a height of sixty cubits.* He said, "Go and greet that group of angels and hear how they return your greeting; that is the method of greeting for you and your progeny." [Adam] said, "Peace be upon you," and they replied, "Peace be upon you and Allah's mercy." Thus, they added "and Allah's mercy." Everyone who enters Paradise will be in Adam's image, and *people have been decreasing until this day.*[1]

Modern concerns surrounding this hadith rest primarily on two issues, one positive (the existence of opposing data) and one negative

1 *Ṣaḥīḥ al-Bukhārī* (Jeddah: Dār Ṭawq al-Najāh, 2001), no. 3326; *Ṣaḥīḥ Muslim* (Jeddah: Dār Ṭawq al-Najāh, 2013), no. 2841; *Jāmiʿ Maʿmar b. Rāshid*, in *Muṣannaf ʿAbd al-Razzāq* (Beirut: al-Majlis al-ʿIlmī, 1982), no. 19435. Abū Zurʿa al-ʿIrāqī (d. 826 AH) writes that the words "until this day" show that the gradual decrease culminated at the time of Prophet Muḥammad (ṣ), after which human height plateaued. See Abū Zurʿa al-ʿIrāqī, *Ṭarh al-tathrīb fī sharḥ al-Taqrīb* (Beirut: Dār Iḥyāʾ al-Turāth al-ʿArabī, n.d.), 8:106–107.

(the lack of supporting data). First, based on our current knowledge of physics and biology, humans could not have been anywhere near ninety feet tall while maintaining the same physiological and anatomical properties and functions.[2] Second, fossil records suggest that the average human height has always remained within the range of six feet.[3] Some have dismissed these concerns as modern or speculative in nature. However, difficulty in reconciling this hadith with empirical evidence is not an entirely new phenomenon. The doyen of latter-day hadith scholarship Ibn Ḥajar al-ʿAsqalānī (d. 852 AH) and the renowned historiographer Ibn Khaldūn (d. 808 AH), among other premodern scholars, expressed similar sentiments.

There are, however, concepts that need to be clarified to better appreciate the nuances involved in examining these hadith. From the methods of conflict resolution, to competing divisions of hadith, to the status of Biblical narratives, this study weaves together multiple disciplines that require a degree of literacy in the Islamic sciences. It will, therefore, be helpful to introduce some of the key terms and ideas employed throughout this work.

From rational to empirical concerns

In the intellectually vibrant milieu of the eighth-century Levant, the Ḥanbalī polymath Ibn Taymiyya (d. 728 AH) recognized that the binary of reason vs. revelation prevalent in the scholastic discourse of his time was unwarranted and superficial. Instead, he proposed

---

2 William H. Press, "Man's Size in Terms of Fundamental Constants," *American Journal of Physics* 48 (1980): 597; Adrien Marck et al., "Are We Reaching the Limits of Homo Sapiens?," *Frontiers in Physiology* 8 (2017): 812; Thomas Samaras, "Human Scaling and the Body Mass Index," in *Human Body Size and the Laws of Scaling: Physiological, Performance, Growth, Longevity and Ecological Ramifications*, ed. Thomas Samaras (New York: Nova Science Publishers, 2007), 17–27.

3 Andrew Gallagher, "Stature, Body Mass, and Brain Size: A Two-million-year Odyssey," *Economics and Human Biology* 11 (2013): 558; Michael Hermanussen, "Stature of Early Europeans," *Hormones* 2, no. 3 (2003): 175–178; K. Mathers and M. Henneberg, "Were We Ever That Big? Gradual Increase in Hominid Body Size Over Time," *Homo* 46, no. 2 (1995); Dennis Styne and Henry McHenry, "The Evolution of Stature in Humans," *Hormone Research* 39 (1993).

that the discussion be reframed from a conflict between the sources themselves to a conflict between the epistemic value of specific evidence from each source, for both sound reason and authentically transmitted revelation cannot reach an impasse.[4] It would not be a stretch to reframe the current debate on the conflict between science and scripture, *mutatis mutandis*, in a similarly nuanced fashion.[5] However, science in this context should not be confused with the concomitant philosophical assumptions that too often blur the lines between neutral scientific inquiry and dogmatic scientism, the worldview that holds science as the ultimate arbiter of defining reality.[6]

After pointing out the theological flaws of the philosophers, Ibn al-Qayyim (d. 751 AH), Ibn Taymiyya's chief acolyte, expresses his frustration with those on the opposite end of the spectrum who reject anything remotely associated with the philosophers. As a result, they stubbornly deny simple facts such as the earth being round or the process of eclipses. Adding insult to injury, they then attribute their lopsided thinking to the messengers of God.[7] One is reminded of al-Ghazālī's coup de grâce against the philosophers, which he himself admits was primarily directed at the Neoplatonic metaphysics espoused and expanded by the *falāsifa* and not at philosophy as a

---

4 See, for instance, Ibn Taymiyya, *Darʾ taʿāruḍ al-ʿaql wa-l-naql* (Riyadh: Jāmiʿat al-Imām, 1991), 1:86–87; Carl Sharif El-Tobgui, *Ibn Taymiyya on Reason and Revelation: A Study of* Darʾ taʿāruḍ al-ʿaql wa-l-naql (Leiden: Brill, 2020), 156–163.

5 For an overview of the different approaches to the topic of Islam and science, see Ibrahim Kalin, "Islam and Science," *Oxford Islamic Studies Online*, http://www.oxfordislamicstudies.com/Public/focus/essay1009_science.html (accessed 5/12/2021).

6 Daniel Dennett admits that "there is no such thing as philosophy-free science, there is only science whose philosophical baggage is taken on board without examination." See Daniel Dennett, *Darwin's Dangerous Idea: Evolution and the Meanings of Life* (New York: Simon & Schuster Paperbacks, 1995), 21. For an overview of scientism and the problems of adopting it from a religious standpoint, see Mikael Stenmark, *Scientism: Science, Ethics and Religion* (New York: Routledge, 2018), chap. 1; Shoaib Ahmed Malik, *Atheism and Islam: A Contemporary Discourse* (Abu Dhabi: Kalam Research & Media, 2018), 23–24.

7 Ibn Qayyim al-Jawziyya, *Miftāḥ dār al-saʿāda* (Jeddah: Majmaʿ al-Fiqh al-Islāmī, 2010), 3:1417–1419.

discipline per se.[8] What is often assumed to be a conflict between scripture and science is rather a conflict between scripture and a particular unspoken metaphysics that takes on the guise of science.

A three-tiered model of conflict resolution

Since the formative period of Islamic intellectual history, Muslim scholars have systematically addressed hadith that conflict with other evidence.[9] Centuries of intense dialogue by hadith experts such as Imam al-Shāfiʿī (d. 204 AH), Ibn Qutayba (d. 276 AH), and al-Ṭaḥāwī (d. 321 AH) with hadith-wary critics like the Muʿtazila have produced a robust framework for dealing with problematic hadith.[10] The nature of the contentions may have changed over the course of history, but the framework retains its utility. Scholars formulated the following four-tiered approach as a general template for conflict resolution:

(1) harmonize (*jamʿ*) both sides through some form of interpretation (*taʾwīl*);[11]

(2) determine if the scriptural evidence is abrogated (*naskh*);

(3) prioritize one side of the conflict (*tarjīḥ*) based on a wide range of considerations;[12] and

8 Abū Ḥāmid al-Ghazālī, *al-Munqidh min al-ḍalāl* (Beirut: Dār al-Kutub al-ʿIlmiyya, 1988), 41–42.

9 The field that addresses contradictory hadith is known as *mukhtalif/mukhtalaf al-ḥadīth*, and the field that addresses hadith conflicting with other evidence (e.g., Qurʾānic, scientific, rational, or empirical) is known as *mushkil al-ḥadīth.* See Abū Shahba, *al-Wasīṭ fī ʿulūm wa-muṣṭalaḥ al-ḥadīth* (Cairo: Dār al-Fikr al-ʿArabī, 2006), 442–443; Usāma Khayyāṭ, *Mukhtalif al-ḥadīth bayna al-muḥaddithīn wa-l-uṣūliyyīn wa-l-fuqahāʾ* (Riyadh: Dār al-Faḍīla, 2001), 25–26.

10 Prominent titles of the genre include al-Shāfiʿī's *Ikhtilāf al-ḥadīth,* Ibn Qutayba's *Taʾwīl mukhtalif al-ḥadīth*, and al-Ṭaḥāwī's *Sharḥ mushkil al-āthār.*

11 Ibn Rajab al-Ḥanbalī, *Fatḥ al-Bārī fī sharḥ Ṣaḥīḥ al-Bukhārī* (Cairo: Maktabat Taḥqīq Dār al-Ḥaramayn, 1996), 5:155–156; Ibn Ḥazm, *al-Iḥkām fī uṣūl al-aḥkām* (Beirut: Dār al-Āfāq al-Jadīda, 1983), 2:21; Abū Jaʿfar al-Ṭaḥāwī, *Sharḥ maʿānī al-āthār* (Beirut: ʿĀlam al-Kutub, 1994), 4:274.

12 Yaḥyā b. Sharaf al-Nawawī, *al-Minhāj fī sharḥ Ṣaḥīḥ Muslim b. al-Ḥajjāj* (Beirut: Dār Iḥyāʾ al-Turāth al-ʿArabī, 1972), 1:35. On the factors that inform a scholar's

(4) suspend judgment on the matter (*tawaqquf*) until further clarity is achieved.[13]

In this study, the process of conflict resolution is three-tiered; abrogation is excluded as a possibility because the scientific and archaeological evidence is not scriptural and the hadith in question is not legal in nature, which is the purview of abrogation.

Probabilistic and definitive knowledge

Before venturing into the process of conflict resolution, the evidentiary weight of all evidence on the matter should be gauged along the epistemic spectrum ranging from probability (*ẓann*) to certainty (*qaṭʿ*) in both transmission (*thubūt*) and semantic import (*dalāla*).[14] This categorization allows one to assign each piece of evidence its

---

decision to prioritize one side of a conflict, see Mujīr al-Khaṭīb, *Maʿrifat madār al-isnād wa-bayān makānatihi fī ʿilm ʿilal al-ḥadīth* (Riyadh: Dār al-Maymān, 2007), 2:87ff.

13 According to ʿAbd al-Majīd al-Turkumānī, although most scholars arrange the steps as outlined above, the sound view in the Ḥanafī legal school is that one should first seek out abrogation if a timeline can be determined, then prioritization, then harmonization, then interpretative abrogation (*naskh ijtihādī*), and finally cancelation (*tasāquṭ*). See ʿAbd al-Majīd al-Turkumānī, *Dirāsāt fī uṣūl al-ḥadīth ʿalā manhaj al-Ḥanafiyya* (Karachi: Maktabat al-Saʿāda, 2009), 499–506. Sharīf Ḥātim al-ʿAwnī paints a different picture of the Ḥanafī school. He contends that the position of the pioneers of Ḥanafī legal theory (e.g., al-Jaṣṣāṣ [d. 370 AH] and al-Bazdawī [d. 482 AH])—contrary to the likes of Ibn al-Humām (d. 861 AH)—conformed to the order of steps outlined by the other legal schools. He argues that that sequence ought to be the sound view in the Ḥanafī school as well. See Ḥātim al-ʿAwnī, *Manāzil ḥall ishkāl al-taʿāruḍ bayna al-nuṣūṣ al-sharʿiyya bayna al-jumhūr wa-l-Ḥanafiyya*, 49.

14 Probability (*ẓann*) is of various shades depending on factors that either strengthen or weaken its epistemic value (e.g., *ghalabat al-ẓann*, or high probability); not every hadith or rational inference is epistemically equal. Depending on how it is defined, certainty (*qaṭʿ*) could also entertain various degrees. See Muḥammad al-Khinn, *al-Qaṭʿī wa-l-ẓannī fī al-thubūt wa-l-dalāla ʿinda al-uṣūliyyīn* (Damascus: Dār al-Kalim al-Ṭayyib, 2007), 69–74, 88–89.

relevant epistemic status and then choose the most appropriate course of action according to the three-tiered model mentioned above.[15]

When assessing the conflict, if evidence from both sides is epistemically equal, then a plausible explanation should be afforded to ease the tension. Otherwise, priority will be given to the side with the epistemically weightier evidence. When all else fails, suspending judgment is the most ideal course of action. The contours of this formula are derived from the Universal Rule (*al-qānūn al-kullī*) vis-à-vis the conflict between reason and revelation espoused by al-Ghazālī, expounded upon by Fakhr al-Dīn al-Rāzī (d. 606 AH), and revisited by Ibn Taymiyya.[16]

Division of hadith reports

Scholars divided hadith reports based on the number of their chains into *mutawātir* (massively transmitted) and *khabar wāḥid* (limited in transmission, and further divided into sub-categories). The former yields epistemologically certain knowledge (*qaṭʿ*), while the latter yields probable knowledge (*ẓann*); each of these two types of hadith has its own role in establishing religious matters, from law to theology. This division of reports is ascribed to legal theorists and theologians. By the late third century AH, hadith scholars began adopting this division though they had instituted a different system

15 Muftī Taqī ʿUthmānī provides a useful summary of the scholarly method of conflict resolution with examples. See Muḥammad Taqī ʿUthmānī, *ʿUlūm al-Qurʾān* (Karachi: Maktabat Dār al-ʿUlūm), 407–419.

16 On the Universal Rule, see El-Tobgui, *Ibn Taymiyya on Reason and Revelation*, 132–137. Muḥammad Zāhid Mughal proposes to reconcile the longstanding conflict between al-Ghazālī and Ibn Taymiyya regarding the Universal Rule by analyzing the former's application of terms like reason and *burhān* (demonstrative argument) in his other works, arguing that the difference between the two thinkers can be boiled down to semantics. See Muḥammad Zāhid, "ʿAql wa-naql kā taʿaruḍ awr taʾwīl key liey Imām Ghazālī kā qānūn kullī," *al-Sharīʿa* 31, no. 5 (2020): 46–52. See also Frank Griffel, "Al-Ghazālī at His Most Rationalist: The Universal Rule for Allegorically Interpreting Revelation (*al-Qānūn al-Kullī fī t-Ta ʾwīl*)," in *Islam and Rationality: The Impact of al-Ghazālī* (Leiden: Brill, 2015), 118–120.

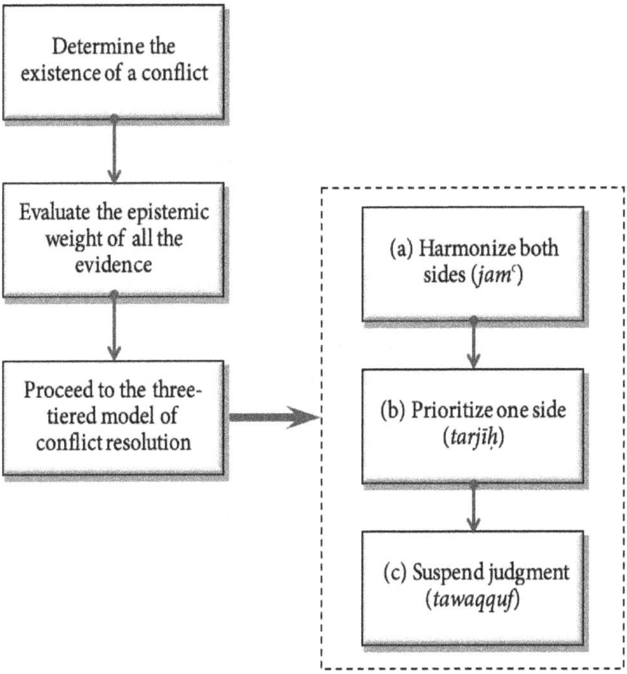

Figure 1: Framework for dealing with scriptural conflict

of analyzing the chains of hadith that focused on the reliability of their transmitters on a spectrum from weak (*ḍaʿīf*) to sound (*ṣaḥīḥ*).[17]

Hadith scholars were nonetheless acquainted with the conversation surrounding the epistemic scale of reports and occasionally utilized a similar vocabulary. Moreover, they were more nuanced in their treatment of solitary reports (*akhbār āḥād*) than their interlocutors gave them credit for.[18] For instance, the leading hadith expert of the third century Abū Ḥātim al-Rāẓī (d. 277 AH) dismissed a

17 Jonathan A.C. Brown, *Hadith: Muḥammad's Legacy in the Medieval and Modern World* (Oxford: Oneworld Publication, 2009), 103–106, 172–182; ʿIṣām ʿĪdū, *Nashʾat ʿilm al-muṣṭalaḥ wa-l-ḥadd al-fāṣil bayna al-mutaqaddimīn wa-l-mutaʾakhkhirīn* (Amman: Arwiqa, 2016), 275ff.

18 See, for instance, Suheil Laher, "Twisted Threads: Genesis, Development and Application of the Term and Concept of Tawatur in Islamic Thought" (PhD diss., Harvard University, 2014), 86.

hadith on judicial law that he acknowledged was a solitary report, while Ibn Khuzayma (d. 311 AH), the author of the famous *Ṣaḥīḥ* collection, rejected the hadith that Adam was created "in the form of the Merciful."[19] Their judgment was based on the fact that the given hadith did not reach a satisfactory level of epistemic certainty for fundamental legal and theological issues.[20]

Evolution of hadith engagement

Muḥyī al-Dīn al-Samarqandī, a contemporary scholar, makes the obvious yet important observation that hadith scholars engaged with, and often critiqued, the content of hadith based on the empirical knowledge that was available to them. The precision of this form of engagement sharpened as their knowledge of the empirical sciences grew. Consequently, modern advancements in science, technology, and archaeology, among other fields, have paved the way for researchers to utilize the framework of premodern scholars with greater accuracy and potentially revisit earlier assumptions in light of more refined findings.[21]

To support this observation, al-Samarqandī quotes al-Shāfiʿī (d. 204 AH), who said that we can determine whether a hadith is true or false "if the transmitter relates what cannot possibly be the case, or what is contradicted by information that is better authenticated and is more indicative of the truth."[22] He further adduces comments

19 Ibn Abī Ḥātim al-Rāzī, *Kitāb al-ʿIlal* (Riyadh: Maṭbaʿat al-Ḥumayḍī, 2006), 4:238–240; Ibn Khuzayma, *al-Tawḥīd* (Riyadh: Dār al-Rushd, 1994), 87.

20 For an insightful analysis on this subject, see Ḥātim al-ʿAwnī, *al-Yaqīnī wa-l-ẓannī min al-akhbār* (Beirut: al-Shabaka al-ʿArabiyya li-l-Abḥāth wa-l-Nashr, 2013), 89ff.; cf. Jonathan A.C. Brown, "Did the Prophet Say It or Not? The Literal, Historical and Effective Truth of *Ḥadīths*," *Journal of the American Oriental Society* 129, no. 2 (2009). There are obvious details and nuances concerning the division and function of reports among hadith scholars, legal theorists, and theologians that require further elaboration.

21 Muḥyī al-Dīn al-Samarqandī, *Naqd matn al-ḥadīth fī ḍaw' natā'ij al-ʿulūm al-tajrībiyya* (Beirut: Dār al-Kutub al-ʿIlmiyya, 2008), 7–10, 25.

22 Muḥammad b. Idrīs al-Shāfiʿī, *al-Risāla* (Baltimore: Johns Hopkins University Press, 1961), ed. and trans. Majid Khadduri, 252.

from Ibn al-Jawzī (d. 597 AH) and Ibn al-Qayyim who regarded a hadith's conflict with sensory perception and observable reality as a sign of its unreliability. Knowledge of these matters can increase with the improvement of the tools and methods used to determine them.[23] The Moroccan polymath Aḥmad al-Ghumārī (d. 1961) penned a treatise entitled *The Conformity of Modern Inventions with That about Which the Leader of Creation Has Informed*, in which he revisits classical explanations of hadith in light of modern advancements.[24]

That some of the proposals to resolve tension surrounding problematic hadith lack precedent—in the present discussion or elsewhere—should not undercut their viability, provided that they operate within the requisite framework founded by classical scholarship. In addition, it should be noted that hadith analysis is not the exclusive domain of hadith scholars. Experts from other disciplines such as law and theology have befittingly engaged in some forms of serious hadith analysis. As Ṭāhir al-Jazā'irī (d. 1920) relates, the expertise of the theologians and legal theorists is superior in some areas of hadith studies, such as addressing hadith that conflict with other texts, while hadith scholars are peerless in their *isnād* analysis.[25] Accordingly, comments about problematic hadith from classical theologians and historians should not be dismissed on the grounds that hadith grading was not their field of expertise.

---

23 Al-Samarqandī, *Naqd matn al-ḥadīth*, 16.

24 Sharaf al-Quḍāh opines that the congruence of a slightly weak hadith or a scriptural interpretation with science elevates its authenticity and validity. See Sharaf al-Quḍāh, "al-Islām wa-l-'ilm fī al-Qur'ān wa-l-sunna," *Majallat Kulliyyat al-Sharī'a* 14 (1996): 35; Jamīl Farīd, *Athar al-'ilm al-tajrībī fī kashf naqd al-ḥadīth al-nabawī* (Beirut: Markaz Namā', 2016), 41–62.

25 Ṭāhir al-Jazā'irī, *Tawjīh al-naẓar ilā uṣūl al-athar* (Aleppo: Maktab al-Maṭbū'āt al-Islāmiyya, 1995), 322. For more on this subject, see Muḥammad 'Awwāma, "Annotations," in *Tadrīb al-rāwī fī sharḥ Taqrīb al-Nawāwī* (Jeddah: Dār al-Minhāj, 2016), 2:141–145; cf. Abū al-Muẓaffar al-Sam'ānī, *Qawāṭi' al-adilla fī al-uṣūl* (Beirut: Dār al-Kutub al-'Ilmiyya, 1999), 1:399.

Isrā'īliyyāt reports

It is important that we define the term *isrā'īliyyāt*, which is frequently mentioned in discussions on the height of Adam.[26] In theory, *isrā'īliyyāt* refer to all forms of pre-Islamic scriptural narratives—Biblical, Talmudic, or otherwise—as well as to reports that were prevalent in Near Eastern antiquity. However, it historically referred to traditions that were of Jewish provenance specifically. At times, *isrā'īliyyāt* reports originated from Biblical or Rabbinic sources but were then modified and reproduced by Muslim transmitters, lacking little resemblance to the original. As such, reports may still be regarded as *isrā'īliyyāt* even when they lack an explicit reference in earlier sources.

The term *isrā'īliyyāt* derives its name from "the Children of Israel (Isrā'īl)," Isrā'īl being the epithet of Prophet Jacob. The term was formally used at a relatively late point in history. The historian al-Mas'ūdī (d. 345 AH) may have been the first to describe traditions of this nature as *isrā'īliyyāt*, although not without a hint of disdain.[27] Apart from direct interactions with Christian and Jewish communities, the major source for *isrā'īliyyāt* was early converts to Islam like the Jewish scholar Ka'b al-Aḥbār (d. 32 AH) and historians like Wahb b. Munabbih (d. 115 AH). The concept is rooted in the prophetic directive permitting the Companions to "narrate from the Children of Israel," with the caveat that the narrated material be historical or pietistic and not conflict with Islamic tenets.[28] Scholars like Ibn Kathīr (d. 774 AH) have decried the excessive reliance on *isrā'īliyyāt*, and this attitude has gained traction in the modern era through the writings of reformist scholars like Rashīd Riḍā (d. 1935).[29]

---

26 A detailed treatment of this subject is found in Jonathan Brown's afterword to this book titled "The Problem of the *Isrā'īliyyāt*."

27 Roberto Tottoli, "Origin and Use of the Term *isrā'īliyyāt* in Muslim Literature," *Arabica* 46, no. 2 (1999).

28 *Ṣaḥīḥ al-Bukhārī*, no. 3461; Shams al-Dīn al-Sakhāwī, *al-Maqāṣid al-ḥasana fī al-aḥādīth al-mushtahira 'alā al-alsina* (Beirut: Dār al-Kitāb al-'Arabī, 1985), no. 396.

29 Muḥammad al-Dhahabī, *al-Isrā'īliyyāt fī al-tafsīr wa-l-ḥadīth* (Cairo: Maktabat Wahba, 1990), 13–15; Musā'id al-Ṭayyār, *al-Taḥrīr fī uṣūl al-tafsīr* (Jeddah: Ma'had al-Imām al-Shāṭibī, 2014), 141–171; Abū Shahba, *al-Isrā'īliyyāt wa-l-mawḍū'āt* (Cairo: Maktabat al-Sunna, 1988); Fahd al-Rūmī, *Manhaj al-madrasa*

Definition of "*dhirā*ʿ"

The hadith in discussion describes the height of Adam as sixty *dhirā*ʿ. Lexically, a *dhirā*ʿ is the distance between one's elbow and the tip of the middle finger, that is, a cubit (lit. elbow in Latin).[30] It has been used to denote different measurements throughout Islamic history, such as a legal cubit (*dhirā*ʿ *sharʿī*), a black cubit, a royal cubit, and a cloth cubit.[31] Commentators differed on which cubit is being referred to in this hadith: Adam's own cubit or the conventional cubit understood by the initial recipients of these words (i.e., the Companions).[32] Ibn al-Tīn (d. 611 AH), the authoritative Tunisian commentator on *Ṣaḥīḥ al-Bukhārī*, explains that the hadith refers to the conventional cubit. If it referred to Adam's own cubit, then his height would have been sixty times the distance between his own elbow and fingertips, thus making his stature disproportionate.[33]

The conventional Islamic usage of a cubit refers to six handbreadths (*qabḍa*: the index finger, middle finger, ring finger, and little finger together), or twenty-four fingerbreadths. This is equivalent to about half a meter, or 1.5 feet.[34] As such, sixty cubits is approximately

---

al-ʿaqliyya al-ḥadītha fī al-tafsīr (Beirut: Muʾassasat al-Risāla, 1983), 312–332; Shari Lowin, "Isrāʾīliyyāt," in *Encyclopaedia of Islam*, 3rd ed., 2019.

30 Murtaḍā al-Zabīdī, *Tāj al-ʿarūs min jawāhir al-Qāmūs* (Kuwait: Wizārat al-Irshād wa-l-Anbāʾ, 1965), 21:5.

31 For various definitions of a *dhirā*ʿ, see Walther Hinz, "Dhirāʿ," in *Encyclopaedia of Islam*, 2nd ed., 1991.

32 Mullā ʿAlī al-Qārī, *Mirqāt al-mafātīḥ sharḥ Mishkāt al-Maṣābīḥ* (Beirut: Dār al-Fikr, 2002), 9:3669. ʿAbd al-Raʾūf al-Munāwī's (d. 1031 AH) presentation of three opinions is a mistake. See ʿAbd al-Raʾūf al-Munāwī, *Fayḍ al-Qadīr sharḥ al-Jāmiʿ al-ṣaghīr* (Cairo: al-Maktaba al-Tijāriyya al-Kubrā), 3:445.

33 Badr al-Dīn al-ʿAynī, *ʿUmdat al-qārī sharḥ Ṣaḥīḥ al-Bukhārī* (Beirut: Dār Iḥyāʾ al-Turāth al-ʿArabī, n.d.), 15:208. In his explanation of the hadith, Ibn Ḥajar paraphrased the above explanation of Ibn al-Tīn in a manner that rendered it faulty. Later scholars then quoted Ibn Ḥajar without scrutiny. See Abū al-Ḥasan al-Sindī, *Ḥāshiyat Ṣaḥīḥ al-Bukhārī* (Lahore: Maktaba Raḥmāniyya, n.d.), 1:586.

34 Muḥammad Shafīʿ ʿUthmānī, *Awzān-e sharʿiyya* (Karachi: Idārat al-Maʿārif, n.d.), 45; al-Nawawī, *al-Minhāj*, 4:195; Ibn Ḥajar al-ʿAsqalānī, *Fatḥ al-Bārī sharḥ Ṣaḥīḥ al-Bukhārī* (Beirut: Dār al-Maʿrifa, 1960), 2:567.

twenty-eight meters, or ninety feet. Figure 2 provides a visual representation of sixty cubits by comparing it with other objects. There is a narration that suggests the hadith is referring to a royal cubit (*dhirāʿ al-malik*), which is seven handbreadths, or thirty fingerbreadths, slightly longer than the conventional cubit.[35] However, as will be explained later, the term "royal cubit" in this narration is unreliable and possibly an insertion (*mudraj*) from one of the transmitters. In terms of its width, Adam's body is described as seven cubits (3.5 meters) in one narration and as six cubits (3 meters) in another, but both of these narrations are weak.[36]

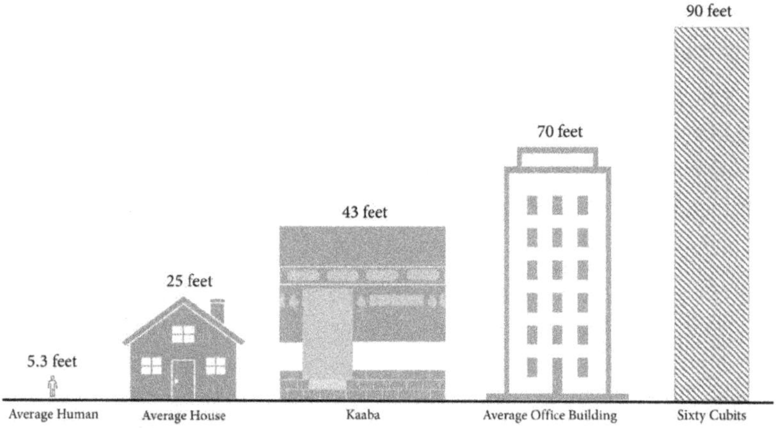

Figure 2: Comparing sixty cubits with the height of various structures

The Baghdadi vizier and jurisconsult Ibn Hubayra (d. 560 AH) explains the relevance of the number sixty in this hadith: "Sixty was chosen because it is an important number that is used in many

35 Muḥammad b. al-Ḥasan al-Shaybānī, *al-Aṣl* (Doha: Wizārat al-Awqāf, 2012), 7:541.

36 See the editors' annotations on Aḥmad b. Ḥanbal, *Musnad Aḥmad*, ed. Shuʿayb al-Arnāʾūṭ et al., 50 vols. (Beirut: Muʾassasat al-Risāla, 1995–2001), no. 7933. The report that mentions his width as six cubits is discussed below under the variants of the sixty cubits hadith.

calculations from which principles are derived and to which [things] are attributed."[37] It appears that Ibn Hubayra is referring to the sexagesimal numeral system that uses the number sixty as a base. The sexagesimal system was used by the ancient Sumerians and later adopted by other civilizations like the Babylonians. It is still in use in some capacity to calculate time and angular measures, e.g., sixty seconds and 360 degrees.[38]

When scripture and empirical data are in conflict, the epistemic weight of evidence from both sides is to be evaluated, both in terms of its transmission and semantic import vis-à-vis scripture and in terms of its veracity vis-à-vis empirical data. The process of conflict resolution then begins by attempting to harmonize both sides. When this is not possible, one side is prioritized. If this is not possible, then judgment is suspended on the matter until further clarity presents itself. For most of Islamic history, the conflict with scripture was based on rational arguments. Given the advancements of empirical science in the modern period, empirical contentions now occupy an unprecedented place in challenging scripture. This methodological shift does not pose a serious threat to scripture because the classical framework for resolving scriptural tension is timeless and adaptable to changing circumstances.

37 Ibn Hubayra, *al-Ifṣāḥ ʿan maʿānī al-Ṣiḥāḥ* (Riyadh: Dār al-Waṭan, 1996), 7:215.

38 On the origins and influence of the sexagesimal system, see François Thureau-Dangin, "Sketch of a History of the Sexagesimal System," *Osiris* 7 (1939); Carol Hill, "Making Sense of the Numbers in Genesis," *Perspectives on Science and Christian Faith* 55, no. 4 (2003); Hildegard Lewy, "Origin and Development of the Sexagesimal System of Numeration," *Journal of the American Oriental Society* 69, no. 1 (1949).

CONFLICT OR CONCORD?

The process of conflict resolution discussed in the preceding chapter yields successful results only when all the evidence on the subject is accurately evaluated. One needs to first determine whether the scriptural evidence is sound. It is futile to expend energy in trying to reconcile, say, an unreliable hadith with empirical realities; if the Prophet (ṣ) did not utter those words, then it is no matter of concern.[1] Reinforcing this idea, Mullā ʿAlī al-Qārī (d. 1014 AH) reminds us of the ancient adage: first stabilize the throne, then engrave it (*thabbit al-ʿarsh thumma 'nqush*).[2] It then needs to be proven that the conflicting empirical data is credible and that an actual conflict exists between it and scripture.

In the case of the hadith on Adam's height, arguments have been put forward that the contentions are subjective, unsubstantiated, and unprecedented. What, then, is the contention? On the one hand, there are authentic hadith stating that Adam and his progeny were

---

1 In his masterpiece dealing with conflicting hadith, al-Ṭaḥāwī sets out to explain the apparent problems and disharmony between reports "transmitted from the Prophet (ṣ) via acceptable chains of transmission related by those of circumspection, honesty, and suitable delivery." See Abū Jaʿfar al-Ṭaḥāwī, *Sharḥ mushkil al-āthār*, ed. Shuʿayb al-Arnāʾūṭ, 16 vols. (Beirut: Muʾassasat al-Risāla, 1994), 1:6. However, it is not uncommon to find scholars expounding on the meanings of hadith that are arguably unreliable. One reason for this practice is that the given scholar did not believe such hadith to be unreliable. He may also have commentated with the hope that if an authentic route of transmission were located later, his commentary would prove helpful. See, for instance, al-Ṭaḥāwī, *Sharḥ maʿānī al-āthār*, 4:331; al-Sakhāwī, *al-Maqāṣid al-ḥasana*, nos. 228 and 934.

2 Mullā ʿAlī al-Qārī, *al-Asrār al-marfūʿa fī al-akhbār al-mawḍūʿa* (Beirut: Dār al-Amāna, 1971), 217.

giants—his progeny gradually decreasing in height—while on the other hand, there is credible scientific and archaeological evidence that undermines that claim. Furthermore, some have attempted to justify the idea of human giants by citing Qur'ānic and post-prophetic evidence. In this chapter, we examine the main scriptural, archaeological, and scientific evidence for and against the apparent claim of the hadith to determine the reality of the conflict.

Scripture

Those who advocate for the existence of giants adduce a plethora of post-prophetic reports that depict earlier nations as giants.[3] Most of these reports, however, are either not reliably transmitted or are based on *isrāʾīliyyāt*.[4] Ibn ʿAbbās, Qatāda b. Diʿāma (d. 118 AH), ʿAṭāʾ b. Abī Rabāḥ (d. 114 AH), and others are reported to have stated that when Adam fell from heaven, he was so tall that he physically reached the sky, after which his height was reduced to sixty cubits.[5] In addition to their defective transmission, these reports were taken from *isrāʾīliyyāt* literature and Near Eastern lore, as noted by the

3 Badr-e ʿĀlam Mīrathī, *Ḥashiyat al-Badr al-sārī*, in *Fayḍ al-Bārī ʿalā Ṣaḥīḥ al-Bukhārī* (Beirut: Dār al-Kutub al-ʿIlmiyya, 2005), 4:342.

4 Farīd Jamīl, *Athar al-ʿilm al-tajrībī*, 138. There are numerous mentions of giants throughout the Hebrew Bible. In the book of Genesis (6:4), we read: "The Nephilim [giants] were on the earth in those days, and also afterward," and in Numbers (13:33), we read: "We saw the Nephilim there [the descendants of Anak come from the Nephilim]. We seemed like grasshoppers in our own eyes, and we looked the same to them." King Og's bedstead was nine cubits in length (Deuteronomy 2:10–11), and Goliath was taller than six cubits (Samuel 17:4). See James L. Kugel, *Traditions of the Bible: A Guide to the Bible as It Was at the Start of the Common Era* (Cambridge, MA: Harvard University Press, 1998), 181–183, 204; Deidre Donnelly and Patrick Morrison, "Hereditary Gigantism: The Biblical Giant Goliath and His Brothers," *Ulster Medical Journal* 83, no. 2 (2014): 86–88.

5 See, for instance, ʿAbd al-Razzāq al-Ṣanʿānī, *Muṣannaf*, nos. 9090, 9096; Jalāl al-Dīn al-Suyūṭī, *al-Durr al-manthūr fī al-tafsīr bi-l-maʾthūr* (Beirut: Dār al-Fikr, 2011), 1:136.

historian Abū al-Ḥasan al-Madā'inī (d. 225 AH).[6] The idea that Adam's height physically reached the sky and later shrunk in size is explicitly mentioned in the Talmud:

> As Rabbi Elazar said: The height of Adam the first man reached from the ground to the skies, as it is stated: "Since the day that God created man upon the earth, and from one end of the heavens" (Deuteronomy 4:32). When he sinned, the Holy One, Blessed be He, placed His hand upon him and diminished him, as it is stated: "You fashioned me behind and before, and laid Your hand upon me" (Psalms 139:5).[7]

There are Qur'ānic verses that demonstrate that certain individuals and civilizations throughout history were relatively larger in stature. However, exegetical comments on these verses often ascribe incredulous descriptions of them that are not rooted in any textual evidence. For example, when Prophet Moses sent spies to scout the Holy Land, these spies returned with information that the area was inhabited by "a domineering people" (*qawman jabbārīn*), as mentioned in the Qur'ān (Q. 5:22). The most apt explanation is provided by the Successor Qatāda, who explained this to mean that "they are larger than we in stature and mightier than we."[8] On the other hand, some have described their height as being such that "seventy of these

6 Ibn al-Jawzī, *al-Muntaẓam fī tārīkh al-mulūk wa-l-umam* (Beirut: Dār al-Kutub al-'Ilmiyya, 1992), 1:202; see also Ṭāriq b. 'Iwaḍ Allāh, *Jāmi' al-masā'il al-ḥadīthiyya* (Cairo: Dār Ibn 'Affān, 2006), 5:143–145.

7 James Kugel traces the origins of this motif in the Bible and highlights the problems with using Psalms 139:5 for that purpose. See Kugel, *Traditions on the Bible*, 82–84. Another Talmudic reference on the height of Prophet Adam reads, "Rabbi Meir says: In the future, the Jewish people will have the stature of two hundred cubits, equivalent to two times the height [*komot*] of Adam the first man, whose height was one hundred cubits." See Bava Batra, 75a. It should be noted that the Hebrew cubit was roughly 17.5 inches, which is slightly shorter than the conventional cubit. See Robert Scott, "The Hebrew Cubit," *Journal of Biblical Literature* 77, no. 3 (1958): 215.

8 'Abd al-Razzāq al-Ṣan'ānī, *Tafsīr 'Abd al-Razzāq* (Riyadh: Maktabat al-Rushd, 1989), 1:188.

spies could take shade under the sole of one of these giants," among other evident exaggerations.[9]

As Ibn Kathīr explains, Qur'ānic verses that state that God increased the stature of certain nations simply mean that they were the largest among the people of their time. The same applies to Ṭālūt (Saul) who was given "great knowledge and stature."[10] Some writers cite the verse containing the words "so you would see the people therein fallen as if they were *hollow trunks of date palms*" to show that earlier civilizations were as tall as date palms.[11] However, the analogy of date palms does not necessarily give an indication of their height. The leading modern exegete Ibn 'Āshūr (d. 1973) explains that after cutting down date palms for general use, the Arabs would lay the trunks on the ground to dry and later use them as pillars. This analogy likens the physical state of the people after being ravaged by destructive winds to trunks of date palms that are scattered on the ground.[12] Furthermore, descriptions of people like 'Ād and Thamūd being larger in stature were mentioned in the Qur'ān as something unique to them, while the hadith in question apparently describes a general phenomenon.[13] Ibn Khaldūn observed that the dwellings of ancient civilizations indicate that they were not drastically tall, as

---

9 For an insightful commentary on this verse, see Rashīd Riḍā, *Tafsīr al-manār* (Cairo: Dār al-Manār, 1948), 6:329–333.

10 Ibn Kathīr, *Tafsīr al-Qur'ān al-'Aẓīm* (Riyadh: Dār Ṭayba, 1999), 1:666, 3:434.

11 Q. 69:7. See, for instance, Muḥammad Zāhid, *Ashraf al-tawḍīḥ: taqrīr urdū Mishkāt al-Maṣābīḥ* (Faisalabad: Maktabat al-'Ārifī, n.d.), 4:324. Date palms average about twenty-three meters in height. See https://www.britannica.com/plant/date-palm (accessed 5/12/2021).

12 Ibn 'Āshūr, *Tafsīr al-Taḥrīr wa-l-tanwīr* (Tunis: al-Dār al-Tūnisiyya, 1984), 29:118. For other explanations of this verse, see Fakhr al-Dīn al-Rāzī, *Mafātīḥ al-ghayb* (Beirut: Dār Iḥyā' al-Turāth al-'Arabī, 1999), 30:622. See also Ibn 'Āshūr's explanation of the verse "as if they were trunks of palm trees uprooted" (Q. 54:20), unrelated to the size of the destroyed people, in Ibn 'Āshūr, *Tafsīr al-Taḥrīr*, 27:194.

13 Muḥammad Taqī 'Uthmānī, *Takmilat Fatḥ al-Mulhim* (Beirut: Dār Iḥyā' al-Turāth al-'Arabī, 2006), 6:158. See also Jeffery Rose's theory on the location of the Land of Aḥqāf and why the people of 'Ād may have been perceived as giants in Jeffery Rose, *Introduction to Human Prehistory in Arabia: The Lost World of the Southern Crescent* (forthcoming 2022), chap. 11.

some have speculated. The claim that they were giants is based on the exaggeration of storytellers.[14]

Archaeology

Archaeologists employ *direct* and *indirect* methods to date human remains. Indirect (or relative) methods study data associated with a given artifact, such as geological information like volcanic ashes near human fossils. This relative dating method provides a general range for when the fossils can be dated, and it is not without its limitations and often subjective techniques. Direct (or absolute) dating methods study the artifact itself to arrive at a more precise date. Examples of direct dating include radiocarbon, uranium series, ESR (electron spin resonance), and AAR (amino acid racemization) dating methods. The scope and margin of error of each method differ. Radiocarbon dating, for instance, is a common method that can date an object back 50,000 years. Uranium series dating, on the other hand, can reach back several hundred thousand years, but there is an assumption that uranium and thorium did not move from the artifact under study; this method, therefore, ultimately dates the uranium in the fossil, thus providing a minimum age.[15]

Fossil records suggest that during the Earlier Upper Paleolithic Era (ca. 40,000–22,000 BP[16]) in Europe, male height averaged 1.7 meters (5.8 feet) and female height averaged 1.6 meters (5.3 feet). This was considered a record high in the trajectory of average human height until the last century.[17] Moreover, fossil records suggest that human height varied over time; it did not decrease (or increase) in

14 Ibn Khaldūn, *Dīwān al-mubtada' wa-l-khabar* (Beirut: Dār al-Fikr, 1988), 1:223; cf. Kashmīrī, *Fayḍ al-Bārī*, 4:342.

15 For a study of direct dating methods as they relate to human fossils, see Rainer Grün, "Direct Dating of Human Fossils," *American Journal of Physical Anthropology* 131, no. 43 (2006).

16 That is, "years before present." The use of BP by archaeologists, geologists, and other scientists refers to radiocarbon ages and to results from other radiometric dating techniques for events before the 1950s.

17 Gallagher, "Stature, Body Mass, and Brain Size," 558; Gert Stulp and Louise Barrett, "Evolutionary Perspectives on Human Height Variation," *Biological*

a linear fashion.[18] It should be noted that the fossil records are often fragmentary—at times, only a skull or a femur is found—or they may not accurately represent every human group alive in a given era.[19] That said, the collective data broadly demonstrates that over thousands of years in different times and places, there was a relatively customary body size, which thus far precludes the notion that our distant ancestors were giants (several times larger than us) who

---

Reviews Cambridge Philosophical Society 91, no. 1 (2016): 209; Marck et al., "Are We Reaching the Limits of Homo Sapiens?," 812.

18 Eva Rosenstock et al., "Human Stature in the Near East and Europe ca. 10,000–1000 BC: Its Spatiotemporal Development in a Bayesian Errors-in-Variables Model," *Archaeological and Anthropological Sciences* 11 (2019): 5676, 5685; İzzet Duyar and Barış Özener, "Evolution of Human Body Height and Its Implications in Ergonomics," *Gaziantep Üniversitesi Sosyal Bilimler Dergisi* 8, no. 1 (2009): 74. Duyar and Özener's study concludes that "when data on human stature is considered as a whole, it revealed that the human body height did not continuously increase in a linear fashion in its evolutionary path but recorded some increases and decreases in different periods of time." John Angel's research on the Eastern Mediterranean region shows that during the Mesolithic period (9000–7000 BC), males averaged 172.5 cm and females 158.7 cm, then during the Early Neolithic (7000–5000 BC), males averaged 169.6 cm and females 155.5 cm, but then during the Bronze Kings (1450 BC), males averaged 172.5 cm and females 160.1 cm, with other height variations throughout a 10,000-year period. See John Angel, "Health as a Crucial Factor in the Changes from Hunting to Developed Farming in the Eastern Mediterranean," in *Paleopathology at the Origins of Agriculture* (Cambridge: Academic Press, 1984), 54–56; cf. Hermanussen, "Stature of Early Europeans," 177.

19 Styne and McHenry, "Evolution of Stature in Humans," 3–6. See, for instance, the recently discovered human fossils from Jebel Irhoud, Morocco, that are dated to about 300,000 years ago in Jean-Jacques Hublin et al., "New Fossils from Jebel Irhoud, Morocco and the Pan-African Origin of Homo Sapiens," *Nature* 546 (2017): 289–292. On the classification of these fossils and others dated around that time, see Rose, *Introduction to Human Prehistory in Arabia,* chap. 7. On how body size is determined by examining fragmentary skeletal remains, see Christopher B. Ruff et al., "Body Mass and Encephalization in Pleistocene *Homo*," *Nature* 387, no. 6629 (1997): 173–176. An increase in the data set and developments in technology and related areas help archaeologists affirm, revise, or refine previous studies. See, for instance, Rosenstock et al., "Human Stature in the Near East and Europe," 5685.

gradually decreased in height. Apart from skeletal remains, a study of prehistoric artifacts and dwellings suggests the same conclusion.[20]

Some writers call to witness discoveries of giant human remains recorded by chroniclers such as the Mamluk-era historian Taqī al-Dīn al-Maqrīzī (d. 845 AH).[21] Apart from the fact that these are second-hand anecdotes, it is difficult to determine the exact nature of these fossils.[22] Adrienne Mayor, a historian who specializes in folk science, writes about tales of unearthing the remains of giants and mythic heroes in Greco-Roman antiquity. Mayor makes a convincing case that the skeletal remains of these supposed giants were from prehistoric mammoths and rhinoceroses that once roamed those fossil-rich lands.[23] In 1577 CE, giant skeletal remains were discovered in Switzerland. An anatomist declared that they belonged to an antediluvian giant. Two centuries later, much to the chagrin of

---

20 On the stone tools used by humans during the Middle Stone Age (300 to 50 thousand years ago) until the Iron Age (2000 to 500 years ago), see John Shea, *Prehistoric Stone Tools of Eastern Africa: A Guide* (New York: Cambridge University Press, 2020), 98–108. For a study of early agricultural communities dating back to 10,000 BCE onward in the Levant, Europe, East Asia, and the Southwest, see Amy Bogaard, "Communities," in *The Cambridge World History: Volume II* (Cambridge: Cambridge University Press, 2015), 124–160.

21 Taqī al-Dīn al-Maqrīzī, *al-Mawā'iẓ wa-l-i'tibār* (Beirut: Dār al-Kutub al-'Ilmiyya, 1997), 1:301; Ibn Taymiyya, *Majmū' fatāwā Shaykh al-Islām Aḥmad b. Taymiyya* (Medina: Majma' al-Malik Fahd, 1995), 27:62. In his famous northern voyage, the Abbasid envoy Aḥmad b. Faḍlān (d. after 310 AH) makes several references to giants. See, for instance, Ibn Faḍlān, *Ibn Faḍlān and the Land of Darkness: Arab Travelers in the Far North*, trans. Paul Lunde and Caroline Stone (London: Penguin Books, 2012), 40, 67, 84.

22 Abū al-Safar al-Kūfī (d. 113 AH) mentions that during a particular battle, they passed through a jungle and came across a man who was sixty cubits tall, but he was headless! See Isḥāq al-Khatlī, *Kitāb al-Dībāj* (Beirut: Dār al-Bashā'ir, 1994), 66–67.

23 Adrienne Mayor, *The First Fossil Hunters: Dinosaurs, Mammoths, and Myth in Greek and Roman Times* (Princeton: Princeton University Press, 2011), 104–129; Adrienne Mayor, *Fossil Legends of the First Americans* (Princeton: Princeton University Press, 2005), chap. 1; Marco Romano and Marco Avanzini, "The Skeletons of Cyclops and Lestrigons: Misinterpretation of Quaternary Vertebrates as Remains of the Mythological Giants," *Historical Biology* 31, no. 2 (2019).

local enthusiasts, the German physician Johann Friedrich Blumenbach (d. 1840) realized upon closer examination that they were the remains of a mammoth.[24] In the late 1800s, the Cardiff Giant began as a lucrative tourist attraction in New York, but within months of its discovery, it became a text-book example of an embarrassing archaeological hoax.[25] More recently, Muḥammad Zaryūḥ cast aspersions on the establishment of archaeological research and championed the idea that there was a systematic conspiracy to suppress evidence of human giants. He writes that in the early 1900s, the Smithsonian Institution destroyed thousands of giant human skeletons to protect the mainstream narrative of human evolution.[26] His source for this event, however, is a satirical website known as World News Daily Report, which by its own admission publishes articles that are "entirely fictional."[27]

---

24 Jan Bondeson, *A Cabinet of Medical Curiosities* (New York: Cornell University Press, 1997), 2. The Irish physician Sir Han Sloan (d. 1753) had debunked giant hoaxes in his time with such frequency that Bondeson gave him the epithet "giant killer." See Bondeson, 85–88.

25 Kenneth Feder, *Frauds, Myths, and Mysteries: Science and Pseudoscience in Archaeology* (New York: McGraw Hill, 2013), 50–62; Mark Rose, "When Giants Roamed the Earth," *Archaeological Institute of America* 58, no. 6 (2005). On a fake image of a giant that gained considerable traction, see James Owen, "'Skeleton of Giant' Is Internet Photo Hoax," *National Geographic*, December 13, 2007, https://www.nationalgeographic.com/news/2007/12/skeleton-giant-photo-hoax/. There were several embarrassing archaeological hoaxes in the past century. A famous example is Piltdown Man, the supposed missing link between man and ape forged by the British archaeologist Charles Dawson (d. 1916). For a recent study on Piltdown Man, see Isabelle De Groote et al., "New Genetic and Morphological Evidence Suggests a Single Hoaxer Created 'Piltdown man,'" *Royal Society Open Science* 3 (2016).

26 Muḥammad Zaryūḥ, *al-Muʿāraḍāt al-fikriyya al-muʿāṣira li-aḥādīth al-Ṣaḥīḥayn* (London: Takwīn li-l-Dirāsāt wa-l-Abḥath, 2020), 3:1432ff.

27 For a scathing critique of Richard Dewhurst's *The Ancient Giants Who Ruled America: The Missing Skeletons and the Great Smithsonian Cover-Up*, see Benjamin Auerbach's review in *American Antiquity* 80, no. 3 (2013).

Science

In the mid-twentieth century, the biologist Florence Moog (d. 1986) criticized the popular satirical novel *Gulliver's Travels* in which the protagonist encounters giants in the fictional land of Brobdingnag. For a myriad of reasons, the existence of these sixty-foot Brobdingnagians, Moog argues, was nothing but a figment of the author's imagination. For one, their 180,000-pound weight could not be held by their frame. Such a skeleton would require shorter legs, a smaller head, a thicker neck, a complete modification of internal organs, and "bones of steel bound by ligaments of wire cables." Moog reminds readers that it is inaccurate to argue for the existence of human giants on the grounds that dinosaurs once existed because the two are so anatomically different that there can be no real comparison.[28] Let us consider the very basic reasons from biology and physics that preclude the possibility of humans reaching a height of ninety feet. For approximation, note that this height is more than twice that of the aptly named Giraffatitan, one of the tallest dinosaurs standing at forty feet tall.[29]

Based on Galileo's square-cube law, if the human body were enlarged while maintaining the same anatomical proportions, the mass would increase by a cubic factor while the cross-sectional area of the muscles and bones would increase by only a square factor. Although this kind of *isometric scaling* applies to uniform geometric objects while biological creatures are subject to *allometric scaling*, which requires a more complex formula, the basic principle remains the same: a striking discrepancy emerges between mass and area at larger sizes requiring a modification of proportions. We can, therefore, use the basic concept of isometric scaling to arrive at a rough approximation of what would happen with a human giant.

---

28 Florence Moog, "Gulliver Was a Bad Scientist," *Scientific American* 179, no. 5 (1948): 52–55. In his influential essay *On Being the Right Size*, John Haldane (d. 1964) begins by highlighting the physiological complications with Pope and Pagan the two giants in John Bunyan's *The Pilgrim's Progress*.

29 Laura Geggel, "What's the World's Largest Dinosaur?" Live Science, January 27, 2019, https://www.livescience.com/34278-worlds-largest-dinosaur.html.

For convenience, we can use Reference Man,[30] who weighs 70 kg and is 170 cm tall (154 lbs. and 5 feet 7 inches). Scaled to 90 feet, Reference Man's mass would increase about 4200 times, thus weighing 294,000 kg (roughly 650,000 lbs.), but his cross-sectional area would increase only 260 times. As a result, he would need to support more than sixteen times more weight for his new cross-sectional area.[31]

A scaling of this magnitude would eventually result in fracturing of the bones, and the muscles would be too weak to continuously sustain the weight.[32] The bones and musculature in the legs would, therefore, have to be disproportionately thicker and wider to support his weight, or he would have to be four-legged.[33] Moreover, the heart would have to be powerful enough to pump blood the height of a three-story building. The required increased pressure in the blood vessels would cause the vessel walls to rupture.[34] The forces of gravity for a ninety-foot-tall human would require a completely different physiological system for venous return. This is partly why dinosaurs had a very different anatomy and physiology to circumvent such constraints.[35] A body of this size would require an entirely different

---

30 A person with the anatomical and physiological characteristics of an average individual that is used in calculations assessing radiation doses. May also be called "standard man."

31 *Report of the Task Group on Reference Man*, ICRP Publication, no. 23, 4 (Oxford: Pergamon Press, 1975). There are obvious limitations and potential modifications to Reference Man. See Wiebke Later et al., "Is the 1975 Reference Man Still a Suitable Reference?," *European Journal of Clinical Nutrition* 64 (2010).

32 Samaras, "Human Scaling and the Body Mass Index," 17–27.

33 John Bonner, *Why Size Matters: From Bacteria to Blue Whales* (Princeton: Princeton University Press, 2006), 34.

34 Venous blood in normal humans returns passively to the heart via the valved veins, which is part of the reason why taller people are more prone to varicose veins. See Lisa Rapaport, "Tall People May Be More Prone to Varicose Veins," Reuters, November 27, 2018, https://www.reuters.com/article/us-health-tallness-varicose-veins/tall-people-may-be-more-prone-to-varicose-veins-idUSKCN1N72CA.

35 On how sauropod dinosaurs were able to maintain blood flow to the brain, see Riley Black, "How Long-Necked Dinosaurs Pumped Blood to Their Brains," *Smithsonian Magazine*, October 21, 2015, https://www.smithsonianmag.com/science-nature/how-long-necked-dinosaurs-pumped-blood-their-brains-180957011/.

anatomy in terms of lungs, nervous system, digestive system, and so on. The ratio between surface area and volume is also important for thermoregulation. Like elephants, a giant human would require a new mechanism to cool down, such as massive ears.[36] The blue whale, the largest animal known to have existed, manages to survive underwater where the force of gravity on the body is counteracted by the force of buoyancy.[37] All of this suggests that if someone were scaled up ninety feet tall, he would be so radically different in his anatomy and physiology that he would hardly be recognizable as a human.

To support a literal reading of the hadith under discussion, some authors cite an article by the contemporary writer ʿIzz al-Dīn Kazābar on the physics and biology related to Adam's height.[38] Although a detailed assessment of his argument is not needed here, it is worth highlighting some of its fundamental flaws. For one, he presents an unreliable hadith that describes Adam's width as seven cubits to suggest that his overall body weight was significantly less than assumed and, therefore, that he would have been able to sustain the weight. He estimates that Adam's weight in that case would have been 125 tons and then asks the reader to consider the *Amphicoelias fragillimus*, a sauropod dinosaur that supposedly weighed 122 tons.[39]

The unreliability of the hadith notwithstanding, this proposal may lower the overall weight, but it renders Adam's body size oddly disproportionate. Although Kazābar calculates the maximum stress that bones can withstand to argue that they could theoretically remain intact under the force of such weight, he neglects the fact that bones are themselves dynamic organs that undergo remodeling at the cellular level due to stress, which would result in marked developmental deformity. He also fails to consider more stressful

---

36 Samaras, "Human Scaling and the Body Mass Index," 20; see also David Esker, "Galileo's Square-Cube Law," Dinosaur Theory, https://dinosaurtheory.com/scaling.html.

37 Bonner, *Why Size Matters*, 16, 31.

38 See, for instance, Zaryūḥ, *al-Muʿāraḍāt al-fikriyya*, 3:1434; ʿIṣām al-Ḥāzimī, *Ṭūl abīnā Ādam ʿalayhi al-salām: shubuhāt wa-rudūd* (Riyadh: Waqf al-Itqān, 2020), 119.

39 ʿIzz al-Dīn Kazābar, "*Ṭūl Ādam wa-l-insān wa-munḥanā nuqṣānihi maʿa al-zamān wa-l-radd ʿalā ʿAdnān.*" December 19, 2012, http://kazaaber.blogspot.com/2012/12/blog-post.html.

activities like falling that would result in the shattering of the bones.[40] In addition, he focuses exclusively on the structural aspects of the legs to the exclusion of other important aspects like the spine. He briefly touches on cardiovascular concerns, but does not address the resulting digestive, vocal, metabolic, respiratory, skin, and nervous system related complications.[41] His analogy with the *Amphicoelias fragillimus* is flawed because the differences between a bipedal human and a quadrupedal herbivorous dinosaur are too significant to allow a meaningful comparison.[42]

Size and longevity

Another argument put forth to support the existence of giants is the correlation between body size and longevity. This line of thinking was introduced by Ibn Qutayba in the third century AH and then expanded by Ibn Hubayra, who posited that because earlier civilizations lived much longer, their bodies would have necessarily been larger.[43] Ibn Hubayra argues that considering the ratio of height to age (60–70 years) for the average person of this *umma*, people from previous civilizations lived for hundreds of years, so it would make sense for them to have been proportionately larger.[44] Studies on the relationship between body size and longevity suggest that the trend *between species* is that larger animals live longer (elephants live longer than mice), but this is not always the case (humans live longer than giraffes). However, *within species* those with smaller bodies live

40 See, for instance, Knut Schmidt-Nielsen, *Scaling: Why Is Animal Size So Important?* (Cambridge: Cambridge University Press, 1984), 6.

41 See, for instance, Moog, "Gulliver Was a Bad Scientist," 52–55.

42 See, for instance, Stephen Brusatte, *Dinosaur Paleobiology* (Hoboken: Wiley-Blackwell, 2012), 146–147, 157–158, 176–182, 208–209.

43 Ibn Qutayba writes that people in the distant past possessed lengthier lifespans and larger bodies. The difference between current human height and human height in the past, he contends, is the same as the difference between the lifespans of both periods. He then mentions that Adam lived for a thousand years. See Ibn Qutayba, *Ta'wīl mukhtalif al-ḥadīth* (Beirut: al-Maktab al-Islāmī, 1999), 407.

44 Ibn Hubayra, *al-Ifṣāḥ*, 7:215.

longer; for instance, shorter humans live longer on average than taller ones.[45] Moreover, longer lifespans do not require a proportionately larger body size. Virtually all peer-reviewed studies on centenarians (i.e., those who lived over one hundred years) show that they are small in size. Therefore, the argument posited by Ibn Hubayra—and reiterated by others—to establish a considerably larger body size based on longevity is not convincing.[46]

The longevity argument is based on the premise that the ancients lived significantly longer than humans today. Current scientific research suggests that the limits of the human lifespan fall within the range of 150 years.[47] Studies conducted on ancient societies challenge the view that earlier humans possessed significantly lengthier lifespans.[48] The scriptural evidence presented to establish a correlation between size and longevity is tenuous. Citing the verse "We sent

---

45 Thomas Samaras et al., "Is Height Related to Longevity?," *Life Science* 72, no. 16 (2003).

46 Thomas Samaras, "How Height is Related to Our Health and Longevity: A Review," *Nutrition and Health* 21, no. 4 (2012); Qime He et al., "Shorter Men Live Longer: Association of Height with Longevity and FOXO3 Genotype in American Men of Japanese Ancestry," *PLoS One* 9, no. 5 (2014).

47 See, for instance, Timothy V. Pyrkov et al., "Longitudinal Analysis of Blood Markers Reveals Progressive Loss of Resilience and Predicts Human Lifespan Limit," *Nature Communications* 12 (2021): 8; Léo R. Belzile et al., "Human Mortality at Extreme Age," *Royal Society Open Science* 8 (2021): 7.

48 Among certain Eastern Mediterranean societies during prehistoric times, the median lifespan was between 29 and 31 years for women and between 31 and 37 years for men. Though very few ancient Egyptians lived over 100 years, the idealized lifespan for them was 110 years. A study on men in ancient Greece and classical Rome between 650 BC and 602 AD reveals that those before 100 BC had a median lifespan of 72 years and those after had a median lifespan of 66 years. More recent research has modified some of these findings, but the differences are negligible for the purposes of our study. See Lawrence Angel, "The Bases of Paleodemography," *American Journal of Physical Anthropology* 30, no. 3 (1969): 430; Sonia Zakrewski, "Life Expectancy," *UCLA Encyclopedia of Egyptology* 1, no. 1 (2015): 9; J. Montagu, "Length of Life in the Ancient World: A Controlled Study," *Journal of the Royal Society of Medicine* 87 (1994); Michael Gurven and Hillard Kaplan, "Longevity Among Hunter-Gatherers: A Cross-Cultural Examination," *Population and Development Review* 33, no. 2 (2007).

Noah to his people, and he lived among them a thousand years save fifty" (Q. 29:14) to establish a correlation between size and longevity is unconvincing; such a deduction is speculative at most.[49] As related by the famous exegete Abū Isḥāq al-Thaʿlabī (d. 427 AH) and argued by some contemporary authors,[50] Noah's lengthy lifespan may have been a miracle specific to him,[51] so the Qurʾānic verse is not indicative of a general trend of extraordinary lifespans for earlier people.[52]

Hadith that state that the average lifespan of a person from this *umma* is "between sixty and seventy" do not specify the lifespans of earlier civilizations.[53] There are a few reports that are relevant to this argument:

---

49 Jihād al-Kandarī and Ashraf al-Quḍāh, "Ḥadīth al-ṣūra: khalaqa Allāh Ādam ʿalā ṣūratihī: dirāsa naqdiyya," *IUG Journal of Islamic Studies* 27, no. 1 (2019): 441, n. 4.

50 Aḥmad al-Thaʿlabī, *al-Kashf wa-l-bayān* (Jeddah: Dār al-Tafsīr, 2015), 3:415. Fakhr al-Dīn al-Rāzī (d. 606 AH) responds to an unnamed physician's criticism of the verse about Noah's lifespan by stating, "This lifespan was not a cause of nature; rather, it was a divine gift (*ʿaṭāʾ ilāhī*)." See al-Rāzī, *Mafātīḥ al-ghayb*, 25:37. I would like to thank Mawlana Tahmid Chowdhury for directing me to al-Rāzī's quote. On the ascription of certain sections of *Mafātīḥ al-ghayb* to al-Rāzī (including the quote cited here), see ʿAbd al-Raḥmān al-Muʿallimī, *Ḥawla tafsīr al-Fakhr al-Rāzī wa-takmilatihi*, in *Majmūʿ rasāʾil al-tafsīr* (Mecca: Dār ʿĀlam al-Fawāʾid, 1424 AH), 305–332; Sohaib Saeed, "Translator's Introduction," in *The Great Exegesis (al-Tafsīr al-kabīr)*, vol. 1, *The Fātiḥa* (Cambridge: The Royal Aal al-Bayt Institute for Islamic Thought and The Islamic Texts Society, 2018), xii–xiii.

51 The Successor Ibn ʿUmayr al-Laythī (d. 74 AH) relates an incident that lends support to the argument that Noah's lifespan was specific to him. Ibn ʿUmayr states that it reached him that Noah was afflicted by his people "waiting generation after generation—each generation worse than the other—until a person said, 'This man lived among our forefathers with the same insanity.'" See Ibn Jarīr al-Ṭabarī, *Jāmiʿ al-bayān fī taʾwīl al-Qurʾān* (Cairo: Dār Hajr, 2001), 12:396. Clearly, this disconnected report cannot serve as definitive evidence, but it demonstrates that the idea that Noah outlived many generations of his people was in circulation within the first century AH.

52 Al-Kandarī and al-Quḍāh, "Ḥadīth al-ṣūra," 441, n. 4.

53 Abū Hurayra narrates that the Prophet (ṣ) said, "The lifespan of my *umma* is between sixty and seventy; few of them will surpass that." See *Jāmiʿ al-Tirmidhī*, nos. 2331, 3550. Another hadith on the authority of Anas states "between fifty and sixty." See Aḥmad b. Ḥanbal, *al-ʿIlal wa-maʿrifat al-rijāl* (Riyadh: Dār al-Khānī, 2001), no. 2231; see also al-Sakhāwī, *al-Maqāṣid al-ḥasana*, no. 130.

- Mālik b. Anas relates via a consecutively broken chain that the Prophet (ṣ) was shown that the lifespans of the ancients were lengthier than the lifespans of his *umma*. This placed them at a disadvantage in terms of the deeds they could perform, so "Allah granted them the Night of Power (*laylat al-qadar*), which is greater than a thousand months."[54] Ibn ʿAbd al-Barr (d. 463 AH) comments on this hadith, "I am unaware of this hadith being reported via a connected chain in any form, nor do I know of any other book that mentions this hadith with a disconnected or connected chain. This is one of the hadith that Mālik alone narrates."[55]

- Ibn ʿUmar narrates that the Prophet (ṣ) sat with his Companions after the ʿaṣr prayer and said, "The difference between your lifespans (*aʿmār*) and the lifespans of those before you is like the time that remains of this day compared to what has passed of it."[56] In addition to some concerns about the chain, a study of all the routes of this and other related hadith show that the message being conveyed is that this *umma* is at the precipice of the Final Hour. In other words, the Prophet (ṣ) is speaking of the "lifespan" of the *umma*, not the lifespans of individual people.[57]

- The hadith about the ascetic who worshiped God for five hundred years has been deemed unreliable by several hadith

54 Mālik, *al-Muwaṭṭa'* (Abū Dhabi: Mu'assasat Zāyid b. Sulṭān, 2004), 3:462.

55 Ibn ʿAbd al-Barr, *al-Tamhīd* (Rabat: Wizārat ʿUmūm al-Awqāf, 1387 AH), 24:373; see also Ibn al-Ṣalāḥ, *Risāla fī waṣl al-balāghāt al-arbaʿa fī al-Muwaṭṭa'* (Beirut: Dār al-Bashā'ir al-Islāmiyya, 2010), 198, 200, 203. Mujāhid relates a report concerning the Night of Power via a disconnected chain which mentions an Israelite who fought in God's path for a thousand months. See al-Bayhaqī, *al-Sunan al-kubrā* (Beirut: Dār al-Kutub al-ʿIlmiyya, 2003), 4:504.

56 *Musnad Aḥmad*, no. 5966.

57 See ʿAbd al-Ra'ūf al-Munāwī, *al-Taysīr bi-sharḥ al-Jāmiʿ al-ṣaghīr* (Riyadh: Maktabat al-Imām al-Shāfiʿī, 1988), 1:360; Sāmī Muḥammad, *al-ʿAmal al-ṣāliḥ*, no. 1278, n. 8; Abū Isḥāq al-Ḥuwaynī, *al-Manīḥa bi-silsilat al-aḥādīth al-ṣaḥīḥa* (Mansoura: Maktabat Ibn ʿAbbās, n.d.), 2:93–98.

experts.[58] Likewise, the story about a tribe from the Israelites that was granted a lifespan of four-hundred years is a non-prophet report from a questionable narrator.[59]

The notion that Adam lived for nearly a thousand years is only narrated in some questionable hadith. This age is mentioned as part of an incident in which Adam offers a portion of his lifespan to another prophet.[60] There are four routes of transmission for this incident, but none of them can serve as a basis for this claim.

- The first route of transmission is via Ibn Abī Dhubāb → Saʿīd al-Maqburī → Abū Hurayra → the Prophet (ṣ). This route mentions that Adam offered sixty years of his lifespan, meaning he lived for 940 years. Al-Tirmidhī comments that this hadith is fair and uncorroborated from this route.[61] Al-Nasāʾī

58 Abū Jaʿfar al-ʿUqaylī, *Kitāb al-Ḍuʿafāʾ al-kabīr* (Beirut: Dār al-Maktaba al-ʿIlmiyya, 1984), 2:144; Shams al-Dīn al-Dhahabī, *Mīzān al-iʿtidāl fī naqd al-rijāl* (Beirut: Dār al-Maʿrifa, 1963), 2:228.

59 This is a post-prophetic statement of Bakr b. Khunays, a questionable narrator. See Ibn Abī al-Dunyā, *Qaṣr al-amal* (Beirut: Dār Ibn Ḥazm, 1997), no. 257; Ibn al-Jawzī, *al-Ḍuʿafāʾ wa-l-matrūkūn* (Beirut: Dār al-Kutub al-ʿIlmiyya, 1986), 1:148.

60 In the Hebrew Bible (Genesis 5:5), Adam is said to have lived for 930 years. For exegetical comments on this passage, see Kugel, *Traditions of the Bible*, 94–95. See Ibn Kathīr's comments on the conflict between the Genesis passage and the hadith cited here; Ibn Kathīr, *al-Bidāya wa-l-nihāya* (Beirut: Dār al-Fikr, 1986), 1:232. The story of Adam offering seventy years of his lifespan to David and subsequently passing away at the age of 930 is recorded in the *Zohar* (1:168b), a mystical commentary on the Torah that is questionably attributed to Rabbi Shimʿon (d. 160 CE). See *The Zohar*, trans. Daniel C. Matt (Redwood City, CA: Stanford University Press, 2005), 3:18. On the authorship of the *Zohar*, see Arthur Green, "Introduction," in op. cit.; Melila Hellner-Eshed, "Zohar," in *Encyclopedia Judaica*, 2nd ed., 2: 656–658.

61 *Jāmiʿ al-Tirmidhī*, no. 3368; see also Abū ʿAbd Allāh al-Ḥākim, *al-Mustadrak ʿalā al-Ṣaḥīḥayn* (Beirut: Dār al-Kutub al-ʿIlmiyya, 1990), 1:32, no. 214. Abū Yaʿlā narrates via Ismāʿīl b. Rāfiʿ [a weak narrator] → al-Maqburī → Abū Hurayra → the Prophet (ṣ) without mention of a thousand years. See Abū Yaʿlā, *Musnad Abī Yaʿlā* (Damascus: Dār al-Maʾmūn li-l-Turāth, 1984), no. 740; on the status of this route, see the editor's footnotes in Ibn Ḥajar al-ʿAsqalānī, *al-Maṭālib al-ʿāliya bi-zawāʾid al-Masānīd al-thamāniya* (Riyadh: Dār al-ʿĀṣima, 2000), 14:230–235.

narrates a truncated version with the same chain and then says that this route is a mistake and that the correct version is Saʿīd al-Maqburī → his father → ʿAbd Allāh b. Salām as his own words (*mawqūf*).[62]

- The second route is via ʿAlī b. Zayd b. Judʿān → Yūsuf b. Mihrān → Ibn ʿAbbās → the Prophet (ṣ). This route mentions that Adam offered forty years and was still granted a full thousand-year lifespan.[63] Ibn Kathīr notes that the hadith of ʿAlī b. Zayd is problematic. He adds that al-Ṭabarānī narrates it via al-Ḥasan as a disconnected report (*mursal*).[64]
- The third route is via Zayd b. Aslam → Abū Ṣāliḥ → Abū Hurayra → the Prophet (ṣ), but Adam's lifespan being a thousand years is not mentioned.[65]
- Ibn Bishrān (d. 430 AH) narrates a fourth route with the inclusion of his age being a thousand years as part of a lengthy report via Abū Maʿshar al-Sindhī → Abū al-Khaṣīb Nāfiʿ and Saʿīd al-Maqburī → Abū Hurayra as his own words (*mawqūf*).[66] Abū Maʿshar is a problematic transmitter and Nāfiʿ is unknown.[67]

Accordingly, there is a dearth of compelling evidence to suggest that Adam lived for nearly a thousand years.

62 Al-Nasāʾī, *al-Sunan al-kubrā* (Muʾassasat al-Risāla, 2001), no. 9976; Aḥmad b. Ḥanbal, *al-ʿIlal*, nos. 5632, 5633.

63 *Musnad Aḥmad*, no. 2270.

64 Ibn Kathīr, *al-Bidāya wa-l-nihāya*, 1:89; Abū al-Qāsim al-Ṭabarānī, *al-Muʿjam al-kabīr* (Cairo: Maktabat Ibn Taymiyya, n.d.), no. 12928.

65 Ibn Saʿd, *al-Ṭabaqāt al-kubrā* (Beirut: Dār Ṣādir, 1968), 1:27; cf. al-Ḥākim, *al-Mustadrak*, no. 3257, where it is graded as authentic. This route does not suffer from the criticisms of the previous routes.

66 Ibn Bishrān, *al-Amālī* (Riyadh: Dār al-Waṭan, 1997), no. 664.

67 Ibn ʿAdī, *al-Kāmil fī ḍuʿafāʾ al-rijāl* (Beirut: Dār al-Kutub al-ʿIlmiyya, 1997), no. 1984; Abū al-Ḥasan al-Dāraquṭnī, *al-Sunan* (Beirut: Muʾassasat al-Risāla, 2004), no. 3529.

The case against the existence of giants rests upon a confluence of evidence. Scriptural evidence in favor of their existence is far too vague, and the relevant post-prophetic reports do not wield the strength to lend them adequate support. Archaeological and scientific data suggest that human height in the past was not substantially greater than the average human height today. The archaeological and scientific arguments to the contrary evidently do not pass muster. Based on the foregoing, it is safe to conclude that there exists a credible conflict between the hadith on the height of Adam and reliable empirical data. The process of resolving this conflict is the focus of this study.

A SURVEY OF PREMODERN VIEWS

Acceptance

A cursory glance at the Islamic scholastic tradition will reveal that most premodern scholars maintained that Prophet Adam's height on earth was literally sixty cubits. By the second century AH, Successors like Ibn Shihāb al-Zuhrī (d. 124 AH) described the height of Adam as sixty cubits without any qualifications.[1] In the following century, Ibn Qutayba quoted an unnamed scholar (*ba'ḍ ahl al-naẓar*) who explained that Allah created Adam in Paradise in his image (*ṣūratihi*) on earth.[2] Al-Nawawī (d. 676 AH) is more explicit. He writes, "This means he was created at the first instance in his image on earth and during his death, and that was with a height of sixty cubits."[3] Based on the hadith that Adam was sixty cubits tall followed by a decrease in human height, Abū Bakr b. al-'Arabī (d. 543 AH) and Ibn al-Qayyim summarily dismissed reports about people who were larger than that height; [4] anything shorter would not have been an issue.

1 Ibn al-Mubārak, *Kitāb al-Raqā'iq* (Bahrain: Wizārat al-'Adl wa-l-Shu'ūn al-Islāmiyya, 2014), 2:771, no. 1661.

2 Ibn Qutayba, *Ta'wīl mukhtalif al-ḥadīth,* 321. For a brief study of Ibn Qutayba's position on the pronoun issue in the hadith "God created Adam in his/His image," see Christopher Melchert, "God Created Adam in His Image," *Journal of Qur'anic Studies* 13, no. 1 (2011): 118–119.

3 Al-Nawawī, *al-Minhāj,* 17:178. Ibn al-Mulaqqin's attribution of this statement to al-Thawrī (d. 161 AH) is either a scribal error or an oversight; the name should be al-Nawawī. See Ibn al-Mulaqqin, *al-Tawḍīḥ li-sharḥ al-Jāmi' al-ṣaḥīḥ* (Damascus: Dār al-Nawādir, 2008), 19:279.

4 Abū Bakr b. al-'Arabī, *Aḥkām al-Qur'ān* (Beirut: Dār al-Kutub al-'Ilmiyya, 2003), 4:391; Ibn al-Qayyim, *al-Manār al-munīf fī al-ṣaḥīḥ wa-l-ḍa'īf* (Aleppo:

After relating the hadith on Adam's height, al-Baghawī (d. 516 AH) quotes a scholar who said that Adam did not go through any physical changes when he came to earth.[5]

Al-Ḥakīm al-Tirmidhī (d. 320 AH) states that the ancients were eighty fathoms (*bāʿ*) tall with lengthy lifespans and their sustenance was massive: "a grain of wheat the size of an ox's kidney, one pomegranate enough for ten people, and a bunch of grapes equal in size."[6] Ibn Hubayra not only believed that Adam was sixty cubits tall on earth, but he also explored the possibility that his horse (that is, his means of transport and other effects) was also proportionately sized.[7] Taqī al-Dīn al-Maqrīzī writes that one should not be quick to dismiss the possibility of giant humans because our intellect and experiences are limited, and to support this he presents the hadith on Adam's height.[8] It is worth adding here that some contemporary scholars like Ḥātim al-ʿAwnī have echoed these sentiments.[9] Al-ʿAwnī writes that Muslims accept matters that cannot be explained by science,

---

Maktab al-Maṭbūʿāt al-Islāmiyya, 1970), 77. For another example of this line of reasoning, see Jalāl al-Dīn al-Suyūṭī, *Muʿtarak al-aqrān fī iʿjāz al-Qurʾān* (Beirut: Dār al-Kutub al-ʿIlmiyya, 1988), 2:82.

5 Muḥyī al-Sunna al-Baghawī, *Sharḥ al-sunna*, ed. Shuʿayb al-Arnāʾūṭ and Zuhayr al-Shāwīsh, 15 vols. (Beirut: al-Maktab al-Islāmī, 1983), 12:255. The positioning of this passage suggests that the quote is from al-Khaṭṭābī (d. 388 AH), but a study of al-Khaṭṭābī's explanation elsewhere shows that his words end before the quote from the unnamed scholar. See Ḥamd al-Khaṭṭābī, *Aʿlām al-ḥadīth fī sharḥ Ṣaḥīḥ al-Bukhārī* (Mecca: Jāmiʿat Umm al-Qurā, 1988), 2228.

6 Al-Ḥakīm al-Tirmidhī, *Nawādir al-uṣūl fī aḥādīth al-Rasūl* (Beirut: Dār al-Jīl, 1992), 1:140.

7 Ibn Hubayra, *al-Ifṣāḥ*, 7:215.

8 Al-Maqrīzī, *al-Mawāʿiẓ wa-l-iʿtibār*, 1:299.

9 Shaykh Muḥammad al-Ḥasan al-Dadaw was asked about the hadith on air, and he offered a brief response touching on the reliability of Hammām b. Munabbih as a transmitter, citing the height decrease among al-ʿAbbās and his progeny and highlighting the limits of archaeology. See https://www.youtube.com/watch?v=cUeqy7839nI (accessed 3/10/2021). In his rejoinder to ʿAdnān Ibrāhīm, Bilāl al-Najjār points out some of Ibrāhīm's methodological flaws and inconsistencies, but al-Najjār ultimately concedes the reliability of the contention and opts for a figurative reading of the hadith. See http://www.aslein.net/showthread.php?t=16214&page=3 (accessed 3/10/2021). In a recent monograph on the subject, ʿIṣām al-Ḥāzimī states that the safest and soundest

such as Prophet Noah's age,[10] so the same logic should be extended to Prophet Adam's height.[11] The exegetical literature is replete with discussions on the large stature of Adam and earlier civilizations.[12] It is true that many of the reports mentioned in these works are based on the *isrāʾīliyyāt*, but they are often cited without contestation, indicating that the authors took no issue with the notion that Adam was extremely tall.[13]

The Meccan poet and historian ʿAbd al-Malik al-ʿĀṣimī (d. 1111 AH) argues that Adam's height was in fact 440 cubits tall. He draws this conclusion based on a number of questionable inferences related to the presumed footprint of Adam located in Sri Lanka (Adam's Peak) and on the assumption that Adam's height was sixty times his own cubit.[14] The Andalusian genealogist Thaʿlaba b. Salāma held the strange view that Adam was two hundred cubits tall as measured by

---

route is to opt for an unwavering acceptance of the hadith in a non-interpretive literal sense. See al-Ḥāzimī, *Ṭūl abīnā Ādam*, 73, 99ff.

10 Note the argument that Prophet Noah's lifespan was a miracle specific to him. See al-Thaʿlabī, *al-Kashf wa-l-bayān*, 3:415; al-Kandarī and al-Quḍāh, "Ḥadīth al-ṣūra," 441, n. 4.

11 See, for instance, Ḥātim al-ʿAwnī, "al-Tamarrud ʿalā al-ʿilm (al-qadīm) bi-daʿwā ʿilm (al-ḥadīth)," al-Madina News, June 13, 2013, https://www.al-madina.com/article/236521; see also https://www.facebook.com/Al3uny/posts/10151593497613953/.

12 See, for instance, *Tafsīr ʿAbd al-Razzāq* (Beirut: Dār al-Kutub al-ʿIlmiyya, 1419 AH), no. 1911; al-Ṭabarī, *Jāmiʿ al-bayān*, 18:603; Abū Manṣūr al-Māturīdī, *Taʾwīlāt ahl al-sunna* (Beirut: Dār al-Kutub al-ʿIlmiyya, 2005), 9:450; al-Thaʿlabī, *al-Kashf wa-l-bayān*, 4:106–107; Abū ʿAbd Allāh al-Qurṭubī, *al-Jāmiʿ li-aḥkām al-Qurʾān* (Cairo: Dār al-Kutub al-Miṣriyya, 1964), 6:388.

13 To be clear, scholars like al-Ṭabarī mentioned that they gathered all available reports as transmitted to them but then consigned the responsibility of analyzing the chains of transmission to the reader. As such, the presence of a problematic report in a book is not indiscriminately a sign of the author's agreement with it. See Ibn Jarīr al-Ṭabarī, *Tārīkh al-rusul wa-l-mulūk* (Beirut: Dār al-Turāth, 1967), 1:7–8; Muḥammad Zāhid al-Kawtharī, *Maqālāt al-Kawtharī* (Cairo: al-Maktaba al-Azhariyya, 1994), 404; Ibn Ḥajar al-ʿAsqalānī, *Lisān al-Mīzān* (Aleppo: Maktab al-Maṭbūʿāt al-Islāmiyya, 2002), 4:125.

14 ʿAbd al-Malik al-ʿĀṣimī, *Samṭ al-nujūm al-ʿawālī fī anbāʾ al-awāʾil wa-l-tawālī* (Beirut: Dār al-Kutub al-ʿIlmiyya, 1998), 1:123.

his own cubit, but when Eve was created from him, he shrunk to a hundred cubits.[15]

Rejection

I have come across three instances of direct or indirect reservations regarding this hadith in the premodern interpretive tradition. First, the enigmatic Sistani historian al-Muṭahhar b. Ṭāhir al-Maqdisī (d. ca. 390 AH) observes that "many Muslims reject the idea that [Adam's] height was sixty cubits because it goes against the norm."[16] Based on a narration from Ubayy b. Ka'b discussed below, al-Maqdisī opines that Adam's height was most accurately that of a *saḥūq* (lit. enormous) date palm, which he explains as anything taller than average human height. The sixty cubits description, he argues, was possibly a transmitter's explanatory insertion.[17] His focus on only one route of the hadith on the topic is irrelevant, because his own comments and the observations that he relates from others pertain to the notion of sixty cubits, not to the specific hadith.

The contemporary scholar 'Iṣām al-Ḥāzimī unconvincingly suggests that the words "many Muslims reject..." were interpolated into al-Maqdisī's *Kitāb al-Bad' wa-l-tārīkh* by the French Orientalist

---

15 Ibn Ḥadīda, *al-Miṣbāḥ al-muḍī fī kitāb al-nabī al-ummī wa-rusulihi ilā mulūk al-arḍ min 'arabī wa-'ajamī* (Beirut: 'Ālam al-Kutub, n.d.), 1:22.

16 Muṭahhar al-Maqdisī, *Kitāb al-Bad' wa-l-tārīkh* (Cairo: Maktabat al-Thaqāfa al-Dīniyya, n.d.), 2:99.

17 Al-Maqdisī, *Kitāb al-Bad' wa-l-tārīkh*, 3:22. For a detailed study on the life and works of al-Maqdisī, see Muḥammad al-Bisṭāmī, "al-Muṭahhar al-Maqdisī wa-manhajuhu al-tārīkhī fī *Kitāb al-Bad' wa-l-tārīkh*," (PhD diss., al-Azhar University, 2008); see also Camilla Adang, *Muslim Writers on Judaism and the Hebrew Bible* (Leiden: Brill, 1996), 48–50. Al-Maqdisī prefaces his stance by explaining that height and age are God's creation, and He can increase them as He wills. Hence, unfamiliarity with prolonged lifespans like that of Noah and the massive stature of previous civilizations is not sufficient reason to reject them. Despite this, he maintains that Adam's height exceeded the average human height—based on the definition of a *saḥūq* date palm—but that it was not sixty cubits, which he believes was possibly a transmitter's insertion.

Clément Huart (d. 1926) who edited the text.[18] Figure 3 shows that the relevant words are found in the manuscript that Huart used for his edition of the work. Therefore, the allegation that he tampered with the text is unfounded.

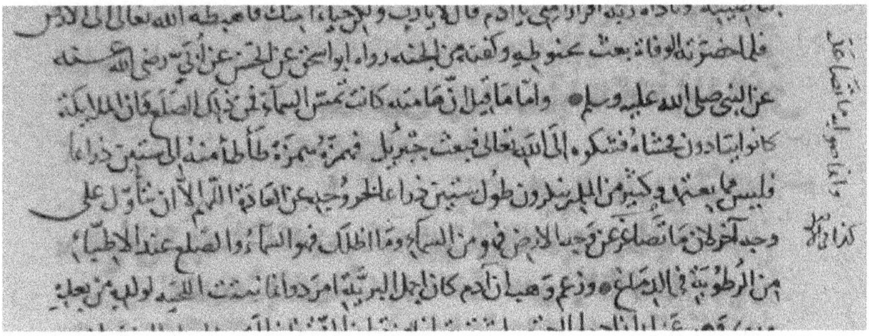

Figure 3: Folio from *Kitāb al-Badʾ wa-l-tārīkh*. MS Istanbul, Süleymaniye – Dāmād Ibrāhīm Pāshā 918, fol. 55r. Courtesy of Süleymaniye Yazma Eser Library

The manuscript evidence renders al-Ḥāzimī's entire case moot.[19] For argument's sake, let us engage his evidence for interpolation. To support his allegation, al-Ḥāzimī says that speaking about "Muslims"

18 Al-Ḥāzimī, *Ṭūl abīnā Ādam*, 92–96. In 1932, ʿAbd al-Raʾūf Dānāpūrī had cast aspersions on the edition of Ibn Saʿd's *Ṭabaqāt* that was edited by the German Orientalist Carl Eduard Sachau (d. 1930). Since the manuscript was edited by an Orientalist, Dānāpūrī claimed that it was impermissible to rely on Sachau's edition to cite Ibn Saʿd unless it was supported by other literature. To verify this claim, the hadith expert Ḥabīb al-Raḥmān al-Aʿẓamī (d. 1992) compared the entire printed book with the original manuscript and concluded that the charge against Sachau was baseless. See Dānāpūrī, *Aṣaḥḥ al-siyar fī hady khayr al-bashar* (Karachi: Aṣaḥḥ al-Maṭābiʿ, n.d.), 7, 15–16; Saʿīd Akbarābādī, *ʿUthmān Dhū al-nūrayn* (Mumbai: Maktabat al-Ḥaqq, 2013), 24; see also Ahmed El Shamsy, *Rediscovering the Islamic Classics* (Princeton: Princeton University Press, 2020), 8–19.

19 Al-Ḥāzimī says that the relevant passage is not found in the Maktabat al-Muṣṭafā copy. However, the copy in reference appears to be a typed version of the printed text without any added significance. See https://al-mostafa.info/data/arabic/depot/gap.php?file=001335-www.al-mostafa.com.pdf. The absence of the missing words—which consist of the sentence under discussion and part of the previous sentence—is indicative of a typographical error because the passage does not read well in their absence. The repetition and placement of

in the third person suggests that one is an outsider. Moreover, in hadith-related discussions, scholars refer to the experts of the field and do not concern themselves with the opinions of lay Muslims.

Al-Ḥāzimī's reasoning is problematic because al-Maqdisī uses phrases like "the Muslims argue" or "the Muslims narrate" throughout his book.[20] Al-Maqdisī's usage of the term "Muslims" is not always a reference to the general masses; rather, he also uses it in reference to the body of scholars. For instance, he contrasts the opinion of the Muʿtazila with the "remainder of the Muslims."[21] Al-Ḥāzimī tries to cast doubt on the ascription of the book *Kitāb al-Bad' wa-l-tārīkh* to al-Maqdisī owing to the poor attestation of the author and his work in historical and biographical literature. It is true that there is a paucity of biographical sources that detail al-Maqdisī's life and works. However, Muḥammad al-Bisṭāmī concludes in his doctoral thesis on al-Maqdīsī that the book is undoubtedly attributable to him.[22]

The second scholar with a dismissive view on Adam being excessively tall is the Ashʿarī theologian Abū Bakr b. Fūrak (d. 406 AH). Ibn Fūrak relates the opinion that the Prophet's statement "God created Adam in *his* form" was in reference to one of the Companions in the gathering.[23] He then explains that the Prophet (ṣ) said this to disprove the notion that Adam's stature and height differed from what was ordinary for humans. He further states that all reports that describe Adam's stature to the contrary come to us from *isrāʾīliyyāt* through the likes of Kaʿb al-Aḥbār and Wahb b. Munabbih.[24]

---

the words "*sittūna dhirāʿan*" in the printed edition before and after the relevant passage may have been the cause of its omission in the typed version.

20 See, for instance, al-Maqdisī, *Kitāb al-Bad' wa-l-tārīkh,* 1:102, 113, 159, 169.

21 See al-Maqdisī, *Kitāb al-Bad' wa-l-tārīkh,* 1:188.

22 Muḥammad al-Bisṭāmī, "al-Muṭahhar al-Maqdisī wa-manhajuhu al-tārīkhī," 256.

23 Daniel Gimaret's critical edition of *Mushkil al-ḥadīth* contains the words "*baʿḍ al-mushāhidīn*" (an attendee) and Mūsā ʿAlī's older edition (1985) contains "*baʿḍ al-mashāhīr*" (a famous figure). Gimaret's edition relied on more manuscripts and the wording found therein is more fitting in this context.

24 Ibn Fūrak, *Mushkil al-ḥadīth wa-bayānuhu* (Beirut: ʿĀlam al-Kutub, 1985), 54–55. The context suggests that the words "*wa-l-fāʾida fī al-khabar. . .*" are from Ibn Fūrak and not "*baʿḍuhum*" quoted before. Throughout the section,

The Ḥanafī hadith commentator Mughlaṭāy b. Qalīj (d. 762 AH) appears to have taken Ibn Fūrak's comments as a negation of Adam's height having reached the sky.[25] However, Ibn Fūrak's words "It is unproven from any other route that Adam's stature differed from this stature [referenced by the Prophet (ṣ)] *such that it substantially exceeded what is customary for humans*" clearly demonstrates that he would not have accepted reports that described Adam's height as anything significantly taller than average human height.[26] Ibn al-Mulaqqin's (d. 804 AH) comment that the hadith on sixty cubits conflicts with Ibn Fūrak's position suggests that he understood Ibn Fūrak's words as a negation of the hadith under discussion.[27]

The third scholar is Ibn Khaldūn, who minces no words in disproving the claim that humans in the distant past were extraordinarily larger than average human height due to the lack of any empirical evidence.[28] In the *Muqaddima*, Ibn Khaldun writes:

> There are also other buildings and monuments (*hayākil*), the history of whose builders, whether ancient or recent, is known to us, and we can be certain that the measurements of their bodies were not excessive. This belief is founded solely upon the tales of storytellers who eagerly tell stories of the people of ʿĀd and Thamūd and the Amalekites. In fact, we find the houses of

---

Ibn Fūrak adds explanatory comments after relating opinions, as he promised in the introduction, often with variations of *afāda*, *fāʾida*, etc. The routes of this hadith will be examined later, but it is worth noting that the narration of Adam's height being sixty cubits has also been transmitted as the words of Kaʿb al-Aḥbār. See al-Ḍiyāʾ al-Maqdisī, *al-Muntaqā min masmūʿāt Marw*, fol. 64r, MS al-Assad National Library, item no. 344 under the category of hadith.

25 Mughlaṭāy b. Qalīj, *al-Talwīḥ sharḥ al-Jāmiʿ al-ṣaḥīḥ* (Damascus: Dār al-Kamāl al-Muttaḥida, 1438 AH, via *Mawsūʿat Ṣaḥīḥ al-Bukhārī*), no. 3326.

26 In his discussion, Mughlaṭāy does not include the italicized words from Ibn Fūrak's quote. Furthermore, when stating that reports to the contrary are based on the *isrāʾīliyyāt*, Ibn Fūrak simply affirms that they mention Adam with a different stature and height; he makes no mention of his height reaching the sky.

27 Ibn al-Mulaqqin, *al-Tawḍīḥ*, 29:13.

28 Ibn Khaldūn, *Dīwān*, 1:222–223. Kashmīrī understood Ibn Khaldūn's comments as a rejection of the hadith about sixty cubits. For Kashmīrī's rejoinder, see *Fayḍ al-Bārī*, 4:342.

> Thamūd still existing at this time in Hegra . . . and it has been observed that those houses are not larger than usual inside, nor in size and height.[29]

In another passage, he reiterates the same point and states, "One should not think, as the common people do, that it was because the ancients had bodies larger than our own. Human beings do not differ in this respect as much as monuments and relics differ."[30] He then critiques the rationalization for the claim that human size and age were greater in the past and steadily decreased by asserting that this is a demonstrably arbitrary view that is disproved by archaeological evidence.[31] Ibn Khaldūn's comments influenced subsequent commentary on the hadith, with some expressing their inability to answer these contentions and others attempting to challenge them.[32]

Ambivalence

Imām Mālik (d. 179 AH) and Ibn Ḥajar al-ʿAsqalānī have been portrayed as premodern scholars who were extremely critical of the present hadith, but this is not an accurate representation of their respective stances. Abū Jaʿfar al-ʿUqaylī (d. 322 AH) records that Mālik vehemently rejected the hadith "God created Adam in His/his image" and proscribed anyone from narrating it.[33] It is true that one of the main hadith on Adam's height begins with these words. However, to say that Mālik rejected all hadith describing Adam's height as sixty cubits requires more evidence because this description is found in other hadith that do not mention "God created Adam in His/his

29 Ibn Khaldūn, *The Muqaddimah: An Introduction to History*, trans. Franz Rosenthal (Princeton: Princeton University Press, 1967), 2:240.

30 Ibn Khaldūn, *Muqaddimah,* 1:357.

31 Ibn Khaldūn, *Muqaddimah,* 1:358–359.

32 Ibn Ḥajar's famous noncommittal view on the gradual decrease of human height was informed by Ibn Khaldūn's contentions. On the other hand, Kashmīrī attempted to challenge them by noting the observable size difference among animals and humans. See Kashmīrī, *Fayḍ al-Bārī*, 4:342.

33 Al-ʿUqaylī, *Kitāb al-Ḍuʿafāʾ al-kabīr*, 2:251.

image."[34] As for Ibn Ḥajar's position, this will be discussed in detail later. Interestingly, when al-Ḥasan al-Baṣrī (d. 110 AH) used to narrate the hadith that Adam's height was sixty cubits, he would interject in between his transmission that "God knows best which cubit [is intended]," thereby suspending judgment on his exact height.[35]

The goal of this short chapter was to survey premodern views on the hadith concerning Adam's height as sixty cubits and the gradual decrease of human height. We can identify three groups of scholars in this respect: (1) the majority, who accepted the hadith at face value; (2) at least three scholars (namely, al-Maqdisī, Ibn Fūrak, and Ibn Khaldūn) who were highly critical of the hadith on account of its implications; and (3) a handful of scholars whose views are ambivalent. Having set the stage for our study, we now proceed with the three-tiered model for resolving the tension surrounding this hadith.

34 On Mālik's rejection of the hadith "God created Adam in His/his image," see al-Dhahabī's comments in *Siyar aʿlām al-nubalāʾ* (Beirut: Muʾassasat al-Risāla, 1985), 8:103–104; al-Jazāʾirī, *Tawjīh al-naẓar*, 65–66.

35 Yaḥyā b. Sallām, *Tafsīr Yaḥyā b. Sallām* (Beirut: Dār al-Kutub al-ʿIlmiyya, 2004), 2:815. Al-Ḥasan al-Baṣrī is said to have expressed hesitation in defining a cubit in other contexts as well. See, for instance, Makkī b. Abī Ṭālib, *al-Hidāya ilā bulūgh al-nihāya* (Sharjah: Jāmiʿat al-Shāriqa, 2008), 2:1361; al-Thaʿlabī, *al-Kashf wa-l-bayān*, 27:313.

PART II

CONFLICT RESOLUTION

HARMONIZATION (*JAMʿ*)

Writing at the turn of the twentieth century in response to the encroachment of modern developments in science, the Grand Mufti of Egypt Bakhīt al-Muṭīʿī (d. 1935) acknowledged the need to reinterpret scripture in the face of definitive astronomical and scientific conflict. However, he advised that this should be done in a reasonable manner that accords with the words and phrases of scripture.[1] Nearly a millennium earlier, in his *Incoherence of the Philosophers*, Abū Ḥāmid al-Ghazālī addressed a problematic addition to a hadith about eclipses, stating, "Even if [the hadith] were sound, it would be easier to interpret it metaphorically rather than to reject matters [of astronomy] that are conclusively true."[2] The Tunisian hadith scholar Abū ʿAbd Allāh al-Māzarī (d. 536 AH) wrote that the Prophet's words do not need confirmation from the physicians (*aṭibbāʾ*), and when in conflict, the objections of the latter should be dismissed. However, if the physicians provide empirical evidence to validate their arguments, scholars "will be required to interpret the Prophet's words in a manner that harmonizes them."[3]

1 Bakhīt al-Muṭīʿī, *Tawfīq al-Raḥmān li-l-tawfīq bayna mā qālahu ʿulamāʾ al-hayʾa wa-bayna mā jāʾa fī al-aḥādīth al-ṣaḥīḥa wa-āyāt al-Qurʾān* (Jeddah: Dār al-Minhāj, 2016), 73. On the rapidly shifting intellectual scene in early twentieth-century Cairo that inspired al-Muṭīʿī to author works on the compatibility of science and Islam, see Junaid Quadri, *Transformations of Tradition: Islamic Law in Colonial Modernity* (Oxford: Oxford University Press, 2021), 112–129.

2 Al-Ghazālī, *Incoherence of the Philosophers,* 7; see also al-Muṭīʿī, *Tawfīq al-Raḥmān*, 70–72; Ibn al-Qayyim, *Miftāḥ dār al-saʿāda*, 3:1422–1425.

3 Abū ʿAbd Allāh al-Māzarī, *al-Muʿlim bi-fawāʾid Muslim* (Tunis: al-Dār al-Tūnisiyya, 1991), 3:170; cf. Ibn al-Khaṭīb, *Muqniʿat al-sāʾil ʿan al-maraḍ al-hāʾil* (Riyadh: Dār al-Amān, 2015), 74.

These authorities hailed from starkly different cultural and educational milieus, but their observations converge on several points that shed light on the process of harmonizing and interpreting scripture (*ta'wīl*),[4] which is commonly defined as diverting a text from its apparent preponderant (*rājiḥ*) sense to its non-preponderant (*marjūḥ*) meaning.[5] For one, not only is revisiting scripture in light of new contentions acceptable, but it can also be the more appropriate route when these contentions are properly substantiated. More importantly, the process of harmonization and interpretation is far from ad hoc. There are principles that govern the legitimate scope and occasion for interpretation, such as avoiding a forced interpretation (*ta'assuf*) that is linguistically implausible, ascertaining the epistemic value of the contentions, and ensuring that the interpretation does not contradict other established evidence.[6] By way of illustration, the hadith "Two months of Eid never fall short (i.e., of thirty days): Ramadan and Dhū al-Ḥijja" prima facie contradicts reality.[7] It is, therefore, interpreted figuratively to mean that these two months will never fall short of spiritual value even though their days may be twenty-nine, which

4 On the parameters of acceptable *ta'wīl* in the realm of theology, see Abū Ḥāmid al-Ghazālī, *Fayṣal al-tafriqa bayna al-Islām wa-l-zandaqa* (Damascus: Muḥammad Bījū, 1993), 47–51; al-Ghazālī, *Qānūn al-ta'wīl* (Damascus: Muḥammad Bījū, 1992), 20–30; Sherman Jackson, *On the Boundaries of Theological Tolerance in Islam: Abū Ḥāmid al-Ghazālī's* Fayṣal al-Tafriqa (Oxford: Oxford University Press, 2002), 43ff. For a comprehensive study of Ibn Taymiyya's understanding of *ta'wīl*, see El-Tobgui, *Ibn Taymiyya on Reason and Revelation*, 179ff.

5 Two other definitions of *ta'wīl* include (1) a straightforward explanation (*tafsīr*) of the apparent sense of a word and (2) "the ultimate reality of that to which the speech pertains" (*ḥaqīqat mā ya'ūlu ilayhi al-kalām*), which Ibn Taymiyya champions as the original understanding of the Companions and early Muslims to the exclusion of the commonly cited definition. See El-Tobgui, *Ibn Taymiyya on Reason and Revelation*, 184, 188.

6 See, for instance, Aron Zysow, *The Economy of Certainty: An Introduction to the Typology of Islamic Legal Theory* (Atlanta: Lockwood Press, 2013), 59ff.; Jonathan A.C. Brown, *Misquoting Muhammad: The Challenge and Choices of Interpreting the Prophet's Legacy* (London: Oneworld Publications, 2014), 83ff.; Khayyāṭ, *Mukhtalif al-ḥadīth*, 125–170.

7 Al-Ṭaḥāwī, *Sharḥ ma'ānī al-āthār*, 2:58.

is a reasonable way to harmonize the hadith with reality.[8] Contrast this with the call to not take the story of Prophet Adam and similar stories in the Qur'ān literally on the grounds that "they are metaphors to learn lessons."[9] This argument, reminiscent of the philosophical theory of "imaginalization" (*takhyīl*),[10] is a prime example of a forced interpretation that asks readers to completely dismiss the apparent reading of scripture without any serious justification.[11]

The apparent meaning of the hadith under study is that (1) Adam was created sixty cubits tall, with no indication that his stature changed, and (2) humankind has since been decreasing in size until they reached what is considered average human height today. Let us examine how scholars revisited this understanding of both parts of the hadith.

---

8 *Jāmiʿ al-Tirmidhī*, no. 692.

9 See, for instance, Rana Dajani, "Evolution and Islam: Is There a Contradiction?," in *Islam and Science: Muslim Responses to Science's Big Questions*, ed. Usama Hasan and Athar Osama (London: Ihsanoglu Task Force, 2016), 146.

10 The theory of *takhyīl* espoused by Muslim Peripatetic philosophers such as al-Fārābī (d. 339 AH) and Ibn Sīnā (d. 427 AH) entailed an interpretation of matters like the afterlife as "a symbolic representation of truths (*takhyīl li-l-ḥaqā'iq*) to benefit the masses, not to actually clarify reality or guide creation to elaborate truths," which Shams al-Dīn al-Saffārīnī (d. 1188 AH) emphatically notes is antithetical to axiomatic Islamic tenets. See al-Saffārīnī, *Lawāmiʿ al-anwār al-bahiyya* (Damascus: Mu'assasat al-Khāfiqayn, 1982), 116; cf. Ibn Taymiyya, *Majmūʿ fatāwā*, 4:67. For a theological critique of *takhyīl* in the thought of al-Fārābī and Ibn Sīnā, see Nazir Khan and Yasir Qadhi, "Human Origins: Theological Conclusions and Empirical Limitations" (Dallas: Yaqeen Institute, 2018), 8–9; El-Tobgui, *Ibn Taymiyya on Reason and Revelation*, 59–56; Deborah Black, "Al-Fārābī," in *The Routledge History of Islamic Philosophy* (New York: Routledge, 1996), 181–182. The Bāṭiniyya were a heterodox sect also known for their extreme allegorical and esoteric readings of scripture. They rejected, inter alia, prophetic miracles and angels and held a convoluted understanding of prophethood. See Abū Ḥāmid al-Ghazālī, *Faḍā'iḥ al-Bāṭiniyya* (Kuwait: Dār al-Kutub al-Thaqāfiyya, n.d.), 40–42; Maḥmūd Shukrī al-Ālūsī, *Rūḥ al-maʿānī* (Beirut: Dār al-Kutub al-ʿIlmiyya, 1994), 8:471.

11 For a critique of this line of reasoning in the conversation on evolution and Islam, see Khan and Qadhi, "Human Origins," 6–12.

Specific to Paradise

With respect to the first clause, the Yemeni hadith expert ʿAbd al-Raḥmān al-Muʿallimī (d. 1966) posits a reasonable interpretation. He writes that the hadith refers to Adam's height in Paradise before he came to the world. Adam then arrived on earth with reduced height appropriate for his earthly existence but taller than the current average human height; subsequent generations decreased in height.[12] Anwar Shāh Kashmīrī (d. 1933), the leading polymath of the Indian subcontinent, proffered a similar answer before al-Muʿallimī. Kashmīrī added that things are relative to their place and time, so just as a day in the afterlife equals a thousand years in this worldly life, Adam's height was accordingly sixty cubits in Paradise where everything is larger.[13]

That the description of sixty cubits refers to Adam's height in Paradise can be inferred from the fact that the relevant hadith mention his height in relation to Paradise. It is incorrect to infer this interpretation from the words "sixty cubits *fī al-samāʾ* (lit. in the sky)" in some routes of the hadith.[14] The words *fī al-samāʾ* do not mean "in Paradise." Rather, this is a figure of speech emphasizing how tall he was. Ibn ʿAṭiyya (d. 542 AH) explains that the Arabs use this phrase to indicate "length in an upward direction," then cites this hadith as an example.[15] A similar phrase was used by the Prophet (ṣ) to describe

---

12 ʿAbd al-Raḥmān al-Muʿallimī, *al-Anwār al-kāshifa li-mā fī Kitāb Aḍwāʾ ʿalā al-sunna min al-zalal wa-l-taḍlīl wa-l-mujāzafa* (Jeddah: Majmaʿ al-Fiqh al-Islāmī, 2012), 259.

13 Kashmīrī, *Fayḍ al-Bārī*, 4:342–343. Before offering this interpretation as a possibility (*yuḥtamal*), Kashmīrī tries to make a case for the decrease of human height by noting the difference in height between different animals and people, like those born in India before British colonization compared to those born afterward or the difference between villagers and city dwellers. He makes this observation when criticizing Ibn Khaldūn's rejection of the hadith under discussion.

14 Though Kashmīrī explains the phrase "*fī al-samāʾ*" as an explanation of height, he then states that it can be a reference to Paradise. See Kashmīrī, *Fayḍ al-Bārī*, 4:342.

15 Ibn ʿAṭiyya, *al-Muḥarrar al-wajīz fī tafsīr al-Kitāb al-ʿAzīz* (Beirut: Dār al-Kutub al-ʿIlmiyya, 1422 AH), 3:335.

Abraham; the Prophet (ṣ) said that he saw a tall man "whose head nearly reached the sky (*fī al-samā'*)."[16]

Resolving a contention

The interpretation of Adam's height being sixty cubits specifically in Paradise is problematized by the last part of the hadith, "*fa-lam yazal al-khalq yanquṣu ba'dahu ḥattā al-ān*," rendered as "thereafter, humankind has been decreasing until this day," indicating a gradual decrease in height.[17] This phrase suggests that Prophet Adam was sixty cubits on earth, and thereafter his progeny continued to decrease in height until the time of Prophet Muḥammad (ṣ).[18]

The renowned Ḥanafī legal expert, Muḥammad Taqī 'Uthmānī, attempts to resolve this objection by arguing that this phrase can be interpreted as "*lam yazal yūladu nāqiṣan* (humankind continues to be born decreased)" to show a static continuation—i.e., humankind has maintained Adam's decreased height relative to his original height in Paradise as sixty cubits. According to this understanding, the hadith does not suggest a gradual decrease from generation to generation.[19]

This explanation may strike readers as compelling, but it has shortcomings. *Lam yazal* is a verb of continuity (*fi'l istimrār*), which indicates the continuance of an action when followed by a verb,

---

16 *Ṣaḥīḥ al-Bukhārī*, no. 7047.

17 Abū Zur'a al-'Irāqī writes that the words "until this day" show that the gradual decrease culminated at the time of the Prophet (ṣ), after which human height plateaued. See al-'Irāqī, *Ṭarh al-tathrīb*, 8:106–107. Ibn al-Qayyim believes that the decrease in height is a result of mankind's sins. See Ibn al-Qayyim, *al-Dā' wa-l-dawā'* (Mecca: Dār 'Ālam al-Fawā'id, 1429 AH), 1:151; for other explanations, see Ibn Hubayra, *al-Ifṣāḥ*, 7:216.

18 This objection does not apply to al-Mu'allimī's explanation because he argues that Adam came to earth taller than ordinary height and then human height decreased gradually in relation to that height. However, this explanation is not clear from the context, given that the idea of a gradual decrease in height is mentioned after the description of sixty cubits. Moreover, this explanation does not resolve the conflict with the data that suggests that human height did not decrease in a linear fashion.

19 'Uthmānī, *Takmilat Fatḥ al-Mulhim*, 6:158.

such as "*yanquṣu*" (to decrease) in our case, implying a continuous decrease. When followed by an active participle (*ism fāʿil*), such as "*nāqiṣ*," *lam yazal* implies the continuance of a state already reached (i.e., they remained [in a] decreased [state]).[20] The present hadith, as it turns out, uses a verb and not an active participle. ʿUthmānī seems to have taken note of this problem, and in order to circumvent it, he considers the verb "*yūladu*" to be implicit in the sentence and substitutes the verb "*yanquṣu*" with the active participle "*nāqiṣ*." Thus, the passage now reads: humankind is continuously born with decreased height [relative to Adam's height in Paradise]. This maneuver may resolve the problem, but the assumption that an entirely different verb is implicit or that the active participle is intended is questionable.[21]

Ibn Baṭṭāl (d. 449 AH) writes that the words "*lam yazal yanquṣu*" mirror the verse "We have created man in the best of stature, then We returned him to the lowest of the low."[22] Man goes through the vicissitudes of life until he reaches perfection, followed by a steady decline. God has kept this deficiency (*naqṣ*) in mankind as proof that if life on this world goes through this process, then the world itself can ultimately come to an end, contrary to the beliefs of the

20 On the linguistic implications of verbs (*fiʿl*), nouns (*ism*), and active participles (*ism faʿil*) in Arabic, see Fāḍil al-Sāmarrāʾī, *Maʿānī al-abniya al-ʿarabiyya* (Amman: Dār ʿAmmār, 2007), 9–16, 41ff.; al-Sāmarrāʾī, *Maʿānī al-naḥw* (Amman: Dār al-Fikr, 2000), 1:240–243.

21 A possible solution to this problem is to suggest that the words "*fa-lam yazal al-khalq yanquṣu*" transmitted by Hammām b. Munabbih were paraphrased (i.e., *riwāya bi-l-maʿnā*) and that the original wording would have accommodated the explanation that humans "remained short" relative to Adam's height in Paradise. However, all the routes of the hadith that include these words contain practically the same wording, and there is no indication that the meaning was simply paraphrased. In addition, the notion of a gradual decrease was expressed by Ibn ʿUmar in another context, as will be mentioned in the next chapter. This indicates that the understanding of a gradual decrease was in circulation within the first century AH, and it seems that this understanding is being conveyed in the hadith in question. At any rate, the concerns with the "static decrease" argument highlighted above may have informed Shaykh Yūnus Jawnpūrī's decision not to opt for such an interpretation and instead to propose that a narrator insertion had taken place.

22 Q. 95:4–5.

naturalists (*dahriyya*).[23] Ibn Baṭṭāl did not expressly interpret the last part as a gradual decrease in height. He appears to take "decrease" to mean the natural decline of life, unrelated to height. This explanation is not entirely clear from the context, especially in consideration of the words "His height was sixty cubits; humankind has continued to decrease after him until this day."

Commenting on the first part of the hadith, "God created Adam in His/his form," Ibn Ḥibbān (d. 354 AH) and al-Khaṭṭābī (d. 388 AH) explain that Adam was created in Paradise fully formed and sixty cubits tall, unlike his progeny who have to go through the process of procreation and then grow to reach perfection—from a drop of semen until reaching their complete height as developed humans.[24] They do not explicitly comment on the last part of the hadith. Ibn al-Tīn (d. 611 AH) may have been the first commentator to explicitly describe these words as a gradual decrease in height, an explanation repeated by subsequent commentators.[25] That said, those who maintained that Adam's height was sixty cubits on earth would naturally accept the idea of a gradual decrease in height to explain current human height.

---

23 Ibn Baṭṭāl, *Sharḥ Ṣaḥīḥ al-Bukhārī*, ed. Abū Tamīm Yāsir Ibrāhīm, 10 vols. (Riyadh: Maktabat al-Rushd, 2003), 9:5.

24 *Ṣaḥīḥ Ibn Ḥibbān* (Beirut: Mu'assasat al-Risāla, 1993), no. 6162; al-Khaṭṭābī, *A'lām al-ḥadīth*, 3:2228.

25 See Ibn al-Tīn's explanation in Ibn Ḥajar, *Fatḥ al-Bārī*, 6:366–367. The explanation of "*yanquṣu*" in this hadith as a gradual decrease of human height was also mentioned by Muẓhir al-Dīn al-Zaydānī (d. 727 AH), Ibn Kathīr (d. 774 AH), and al-Kirmānī (d. 786 AH). See Muẓhir al-Dīn al-Zaydānī, *al-Mafātīḥ fī sharḥ al-Maṣābīḥ* (Kuwait: Dār al-Nawādir, 2012), 5:120; Ibn Kathīr, *al-Bidāya wa-l-nihāya*, 1:114; Shams al-Dīn al-Kirmānī, *al-Kawākib al-darārī fī sharḥ Ṣaḥīḥ al-Bukhārī* (Beirut: Dār Iḥyā' al-Turāth al-'Arabī, 1937), 22:73. Though he is widely cited for his hadith commentary, there is scarce information on the life of Ibn al-Tīn. His name is 'Abd al-Wāḥid b. 'Umar, a Mālikī hadith expert who hailed from Sfax, Tunis where he passed away in 611 AH. He is most noted for his commentary on *Ṣaḥīḥ al-Bukhārī* entitled *al-Mukhbir al-faṣīḥ al-jāmi' li-fawā'id Musnad al-Bukhārī al-Ṣaḥīḥ*, which is partially extant in manuscript form. See Nūra al-'Īd, "al-Imām Ibn al-Tīn al-Ṣafāqusī wa-manhajuhu fī al-ḥadīth al-mushkil," *al-Ādāb* 16 (2020).

Yūnus Jawnpūrī (d. 2017), one of the greatest commentators of *Ṣaḥīḥ al-Bukhārī* in recent memory, agrees with Kashmīrī's interpretation of Adam's height and then attempts to resolve the concerns surrounding the idea of a gradual decrease in height. He argues that the phrase "*fa-lam yazal al-khalq yanquṣu baʿdahu ḥattā al-ān*" is a narrator insertion (*mudraj*) into the text of the prophetic hadith.[26] This understanding allows us to interpret Adam's height of sixty cubits as something specific to Paradise without needing to address the concerns of a gradual decrease in height because those words are not part of the prophetic hadith to begin with. Jawnpūrī's argument will be discussed in more detail in the next chapter.

Sixty as hyperbole

A second interpretation of the hadith was put forward by Muḥammad Abū Shahba (d. 1982), who posits that the word "sixty" in this hadith was possibly mentioned to convey hyperbole (*mubālagha*), i.e., Adam was very tall but not specifically sixty cubits.[27] The same use of hyperbole is found in English, as in phrases like "His bag weighs a ton" or "I tried a million times." Although this practice was common in Arabic, it was generally used in conjunction with multiples of the number seven (e.g., 7, 70, 700), like the verse "If you should ask forgiveness for them seventy times, never will Allah forgive them."[28] Occasionally, the number one thousand was also

---

26 It is worth noting that Jaʿfar al-Firyābī (d. 301 AH) narrates the hadith without any mention of the phrase "Everyone who enters Paradise will be in Adam's image, and people have been decreasing until this day." These words are soundly transmitted from the common link, Hammām b. Munabbih, in multiple sources; therefore, it was al-Firyābī who likely truncated the hadith by omitting the passage about gradual human decrease. See Jaʿfar al-Firyābī, *Kitāb al-Qadar* (Riyadh: Aḍwāʾ al-Salaf, 1997), 32, no. 3. On al-Bukhārī's methodology of dealing with narrator insertions in his *Ṣaḥīḥ*, see Wasīm al-Shūlī, "Manhaj al-Bukhārī fī al-ḥadīth al-mudraj," *al-Majalla al-Dawliyya* 6, no. 7 (2016).

27 Abū Shahba, *Difāʿ ʿan al-sunna wa-radd shubah al-mustashriqīn wa-l-kuttāb al-muʿāṣirīn* (Cairo: Maktabat al-Sunna, 1989), 131.

28 Q. 9:80. On why the number seven is used for hyperbole, see Abū al-Suʿūd Efendī, *Irshād al-ʿaql al-salīm ilā mazāyā al-Kitāb al-Karīm* (Beirut: Dār Iḥyāʾ al-Turāth al-ʿArabī, n.d.), 4:87; ʿIṣām al-Dīn al-Qūnawī, *Ḥāshiyat al-Qūnawī ʿalā*

used, such as the verse "on a day the extent of which is a thousand years of those which you count."[29] In another context, Abū al-'Abbās al-Qurṭubī (d. 656 AH) quotes a commentator who mentions that the number one hundred conveys hyperbole. Al-Qurṭubī then objects to this statement because the Arabs customarily conveyed hyperbole with the number seventy, not one hundred.[30]

Interestingly, Shams al-Dīn al-Kirmānī (d. 786 AH) writes that six is a perfect number because it equals the sum of its positive divisors (i.e., a sixth, a third, and half of six together equal six). By multiplying six by ten (which results in sixty), hyperbole can be conveyed.[31] Badr al-Dīn al-'Aynī (d. 855 AH) reiterates the same point without critique.[32] This conversation, however, takes place in reference to the hadith that the branches of faith are "about sixty or seventy." It makes sense to interpret sixty as abundance in this context given the various opinions on what the branches of faith are. Furthermore, many routes of the hadith on the branches of faith contain only the number seventy or they contain both numbers. Most commentators, however, focus on the number seventy when discussing the possibility of hyperbole.[33]

Al-Kirmānī and al-'Aynī do not seem to be forming a universal linguistic maxim; doing so would require more evidence, such as lexical attestation or poetic precedent. Commentators often resort to explaining certain numbers as an expression of hyperbole when the routes of a hadith vary with regard to the given number, but this does not mean that these numbers linguistically carry the meaning of hyperbole in and of themselves (as do multiples of the number

---

Tafsīr al-Bayḍāwī (Beirut: Dār al-Kutub al-'Ilmiyya, 2001), 9:296; cf. Lawrence Conrad, "Seven and the Tasbi': On the Implications of Numerical Symbolism for the Study of Medieval Islamic History," *Journal of the Economic and Social History of the Orient* 31 (1988).

29 Q. 32:5. See Ibn 'Āshūr, *Tafsīr al-Taḥrīr*, 21:24.

30 Abū al-'Abbās al-Qurṭubī, *al-Mufhim li-mā ashkala min talkhiṣ kitāb Muslim* (Beirut: Dār Ibn Kathīr, 1996), 7:83.

31 Al-Kirmānī, *al-Kawākib al-darārī*, 1:84.

32 Al-'Aynī, *'Umdat al-qārī*, 1:127.

33 Ibn Ḥajar, *Fatḥ al-Bārī*, 1:52; al-Nawawī, *al-Minhāj*, 2:3; Sharaf al-Dīn al-Ṭībī, *al-Kāshif 'an ḥaqā'iq al-sunan* (Riyadh: Maktabat al-Bāz, 1997), 2:439.

seven).[34] In the hadith under discussion, there are no reliably narrated alternatives to the number sixty. Reports suggesting that Adam's height was twelve, eighteen, or seventy cubits are unreliable.[35] Hence, it is difficult to explain the hadith as hyperbole.

Other interpretations

Thus far, we have explored two interpretations of this hadith: (1) sixty cubits was Adam's height specifically in Paradise and (2) the number sixty is an expression of hyperbole. There are other, less obvious interpretations.[36] 'Aṭiyya Muḥammad Saqr (d. 2006), who once served as head of al-Azhar's fatwa committee, argues that *dhirāʿ* (cubit) can be measured as five centimeters. A height of sixty cubits, therefore, would mean that Adam was three meters tall (about ten feet).[37] However, he provides no justification for this usage, nor does the application of this isolated definition to the present hadith find any support in scholarly commentary. The Indian scholar and political activist 'Ubayd Allāh Sindhī (d. 1944) provides a thought-provoking interpretation of the hadith. He believed that it should be interpreted as a reference to Adam's height in the Sufi-cum-philosophical realm of similitude (*ʿālam al-mithāl*)—an intermediary nexus between the realms of the seen and the unseen—and not in our material world.[38]

34 For instance, a hadith states that a time will come when there will be one caretaker for every fifty women. One version of this hadith contains the words forty women. Ibn Ḥajar comments that the numbers are possibly used here to express hyperbole, not these specific numbers. See Ibn Ḥajar, *Fatḥ al-Bārī*, 6:334, 9:330.

35 These reports will be discussed in the next chapter.

36 Abū Muslim al-ʿArābilī, a contemporary preacher, puts forward the poorly argued proposition that *ṭūluhu* (lit. his height) refers to Adam's strength and capability and that sixty cubits refers to the six directions. See https://youtu.be/5u8ehKEomAQ (accessed 4/16/2021). On the height of Adam in Shīʿī literature, see Muḥammad Bāqir al-Majlisī, *Biḥār al-anwār* (Beirut: Dār Iḥyāʾ al-Turāth al-ʿArabī, n.d.), 11:126–129.

37 *Kitāb Fatāwā Dār al-Iftāʾ al-Miṣriyya*, 8:123 (https://al-maktaba.org/book/432/3716#p3; accessed 4/16/2021).

38 See Muḥammad 'Abd al-Razzāq Ḥamza's addendum to al-Muʿallimī, *al-Qāʾid ilā taṣḥīḥ al-ʿaqāʾid* (Beirut: al-Maktab al-Islāmī, 1984), 255. Sindhī advocated, as

Scholar	Height of Sixty Cubits	Subsequent Decrease
Anwar Shāh Kashmīrī	Specific to Paradise	—
ʿUbayd Allāh Sindhī	Height in *ʿālam al-mithāl*	—
ʿAbd al-Raḥmān al-Muʿallimī	Specific to Paradise	Minimal decrease
Abū Shahba	"Sixty" expresses hyperbole	—
ʿAṭiyya Saqr	Sixty *dhirā*ʿ amounts to ten feet	—
Yūnus Jawnpūrī	Specific to Paradise	Narrator insertion
Taqī ʿUthmānī	Specific to Paradise	Relatively shorter

Figure 4: Modern interpretive views on the hadith

did Shāh Walī Allāh before him, the interpretation of numerous Qurʾānic verses and hadith through the lens of the realm of similitude—functionally different from Plato's realm of ideas—to resolve contentions surrounding them. See Sindhī, *Sharḥ Ḥujjat Allāh* (Lahore: Maktabat Bayt al-Ḥikma, 1950), 98–103; for a critique of this method, see al-Kawtharī, *Ḥusn al-taqāḍī* (Amman: Dār al-Fatḥ, 2017), 248–249. Sindhī's interpretation of Adam's height is best understood in light of his stance on the location of Paradise and Shāh Walī Allāh's complex depiction of Adam's creation and stages of development as seen from the lens of the realm of similitude. See Ghulām Muṣṭafā Qāsimī, "Introduction," in Dihlawī, *Taʾwīl al-aḥādīth fī rumūz qaṣas al-anbiyāʾ* (Hyderabad: al-Maṭbaʿ al-Ḥaydarī, 1966), 8–14. On the realm of similitude, see Sindhī, 93–98; Fuad Naeem, "The Imaginal World (*ʿĀlam al-Mithāl*) in the Philosophy of Shāh Walī Allāh Dihlawī," *Islamic Studies* 4, no. 3 (2006); ʿAbd al-Fattāḥ al-Yāfiʿī, *ʿĀlam al-mithāl: Ḥaqīqatuhu wa-adillatuhu* (Sanaa: Dār Uṣūl al-Dīn, 2018), 6–8.

As Aḥmad al-Muzanī (d. 356 AH) reminds us, unscrupulously dismissing a problematic hadith is a simple task that even an ignoramus can accomplish. The brilliance of an intellectual shines when he can resolve the contention and provide a suitable interpretation.[39] However, once a scholar exhausts the list of reasonable interpretations and is left unconvinced of their merit, he should proceed to the next step of conflict resolution: prioritizing one side of the conflict.

39 Abū Bakr al-Kalābādhī, *Baḥr al-fawā'id* (Cairo: Dār al-Salām, 2008), 1:540.

PRIORITIZATION (*TARJĪḤ*)

It can be tempting to harmonize conflicting evidence at any cost to avoid the taxing process of prioritizing one side of the conflict, but there are limits to the scope and function of reconciliation. After citing Abū Bakr al-Ismāʿīlī's (d. 370 AH) critique of a hadith in *Ṣaḥīḥ al-Bukhārī*, Ṭāhir al-Jazāʾirī (d. 1920) poses a hypothetical comment: namely, these critiques can be answered by deploying a figurative interpretation. To this he responds: "There is no issue with a reasonable interpretation, but an unreasonable one is not considered. If this gate were to be opened, then any statement could be interpreted contrary to its correct indication."[1] Commenting on the conflicting reports on the Prophet's ascension (*isrāʾ*), Ibn al-Qayyim (d. 751 AH) emphasizes that it only occurred once and disagrees with those who interpret them as multiple occurrences. He then writes that it constitutes poor methodology to forcefully harmonize conflicting iterations of a single hadith instead of prioritizing one iteration and attributing a flaw to one of the transmitters of another iteration.[2]

The conflict between the implications of a hadith and empirical evidence often demanded that scholars prioritize one side of the conflict. For Zayn al-Dīn al-ʿIrāqī (d. 806 AH), the hadith that a Muslim who reaches forty will be protected from leprosy was defective on account of a narrator flaw. However, the observable reality that a sixty-year-old had contracted leprosy left him no choice but to

1 Al-Jazāʾirī, *Tawjīh al-naẓar*, 322.

2 Ibn al-Qayyim, *Zād al-maʿād fī hady khayr al-ʿibād* (Beirut: Muʾassasat al-Risāla, 1994), 2:273, 3:38. For an extensive study on employing "multiple occurrences" as an interpretive tool for conflicting hadith, see Ḥamza al-Bakrī, *Taʿaddud al-ḥāditha fī riwāyāt al-ḥadīth al-nabawī* (Amman: Arwiqa, 2013).

declare it a definite forgery (*mawḍūʿ qaṭʿan*).[3] Ibn al-Mubārak (d. 181 AH) could not accept the hadith that lentils were sacralized on the tongues of seventy prophets. That lentils "are harmful and cause bloating" was the straw that broke the camel's back.[4]

Analyzing the transmission of "sixty cubits"

Hadith are complex narratives often relayed through multiples lines of transmission. A detailed analysis of each route of a hadith's transmission is required to identify its urtext (*aṣl*) and unearth transmission errors.[5] As Ibn al-Madīnī (d. 234 AH) explains, until all the routes of a hadith are collected and analyzed as a whole, its errors will not become apparent.[6] In his study on the influence of empirical science on hadith criticism, the Jordanian scholar Abū Sāra Jamīl Farīd

---

3 Ibn Ḥajar al-ʿAsqalānī, *al-Qawl al-musaddad fī al-dhabb ʿan al-Musnad li-l-Imām Aḥmad* (Cairo: Maktabat Ibn Taymiyya, 1981), 9, 23–24. Ibn Ḥajar's subsequent rejoinder need not detain us at this point because we are not concerned with the grading of this specific hadith. Rather, what is instructive for our purposes is the observable conflict that informed al-ʿIrāqī's decision to regard it as a forgery.

4 Abū Isḥāq al-Jūzajānī, *Aḥwāl al-rijāl* (Faisalabad: Hadith Academy, n.d.), no. 385; cf. Brown, "How We Know Early Ḥadīth Critics Did *Matn* Criticism," 160. For other examples, see Ibn al-Qayyim, *al-Manār al-munīf*, 51–54; Jonathan A.C. Brown, "The Rules of *Matn* Criticism: There Are No Rules," *Islamic Law and Society* 19 (2012): 378.

5 On how hadith scholars engaged with a hadith's urtext and its permutations, see Brown, "Did the Prophet Say It or Not?," 273–275. For a fascinating study that reconstructs the urtext of a hadith compilation, see Ahmed El Shamsy, "The Ur-*Muwaṭṭaʾ* and Its Recensions," *Islamic Law and Society* (published online ahead of print, 2021).

6 Al-Khaṭīb al-Baghdādī, *al-Jāmiʿ li-akhlāq al-rāwī wa-ādāb al-sāmiʿ* (Riyadh: Maktabat al-Maʿārif, 1983), 2:212, no. 1640. Ibn Taymiyya and Ibn al-Qayyim dismissed a particular route of a hadith recorded in *Ṣaḥīḥ Muslim* that describes the biological process of how an infant's *gender* is determined. They gave preference to another route of the same hadith that instead mentions an infant's *appearance* (*shabah*). By opting for the second route, they were able to resolve the contentions leveled against the hadith. See Ibn al-Qayyim, *Tuḥfat al-mawdūd bi-aḥkām al-mawlūd* (Damascus: Dār al-Bayān, 1971), 1:279–281; Ibn al-Qayyim, *Miftāḥ dār al-saʿāda*, 2:737–738.

posits that one way to resolve the contentions surrounding the hadith on Prophet Adam's height is to analyze its various routes and then determine how it fares epistemically when compared to the empirical objections. The following analysis builds on Farīd's study by adding a substantial number of unexplored routes of transmission, employing visual diagrams, and revisiting the strength of his conclusion. The description of Adam's height as sixty cubits is featured in all major hadith collections as part of lengthier hadith on the authority of Abū Hurayra and other Companions. There are four main hadith on the subject, each transmitted through varying routes of transmission. To provide context, select routes of these hadith have been reproduced in their original Arabic in the Appendix.

Hadith 1: The first batch to enter Paradise

The first hadith describes the first batch of people to enter Paradise. Abū Hurayra narrates that the Prophet (ṣ) said, "The first batch will enter Paradise in the image of the full moon . . . in the image (*khalq*) of one man,[7] the feature of their father Adam, sixty cubits in the sky." Abū Zur'a[8] and Abū Ṣāliḥ[9] narrate this hadith from Abū Hurayra with the words "the feature of their father Adam, *sixty cubits in the sky*." On the other hand, al-A'raj,[10] 'Abd al-Raḥmān b. Abī 'Amra,[11]

7 After transmitting this hadith, Muslim states that Ibn Abī Shayba vowelized the word as "*khuluq*," which would mean "with the character of one man."

8 *Ṣaḥīḥ al-Bukhārī*, no. 3327.

9 *Ṣaḥīḥ Muslim*, no. 2834. *Ṣaḥīḥ Muslim*, no. 2834. Muslim narrates this route via Abū Mu'āwiya → al-A'mash → Abū Ṣāliḥ → Abū Hurayra. Though al-A'mash is a *mudallis* and he utilizes the participle *'an* in this chain, al-Dhahabī maintains that his transmission from the likes of Abū Ṣāliḥ should be treated as continuous (*ittiṣāl*). See al-Dhahabī, *Mīzān al-i'tidāl*, 2:224. The corroborating route of 'Umāra from Abū Ṣāliḥ is erroneous according to Imam Aḥmad. See Ibn Rajab al-Ḥanbalī, *Sharḥ 'Ilal al-Tirmidhī* (Zarqa: Maktabat al-Manār, 1987), 2:859–860; al-Maqdisī, *Aṭrāf al-gharā'ib wa-l-afrād* (Beirut: Dār al-Kutub al-'Ilmiyya, 1997), no. 5739.

10 *Ṣaḥīḥ al-Bukhārī*, no. 3246.

11 *Ṣaḥīḥ al-Bukhārī*, no. 3254.

Hammām b. Munabbih,[12] Muḥammad b. Sīrīn,[13] Abū Salama,[14] ʿIyāḍ b. Dīnār,[15] ʿUqba b. Abī al-Ḥasnāʾ,[16] Ziyād al-Makhzūmī,[17] and ʿAṭāʾ b. Yasār[18] narrate a truncated version of the hadith without the words in question. A similar hadith is transmitted from other Companions, such as Abū Saʿīd al-Khudrī,[19] Ibn Masʿūd,[20] Jābir,[21] and Anas[22] without any mention of sixty cubits. Figure 5 outlines the routes of Hadith 1, highlighting the inclusion and exclusion of the relevant words.

Hadith 2: Adam greeting the angels

The second hadith describes the creation of Adam and his greetings to the angels. This is the only version that includes the addition of humankind gradually decreasing in height. Hammām b. Munabbih narrates from Abū Hurayra that the Prophet (ṣ) said:

> Allah created Adam with a height of sixty cubits. He said, "Go and greet that group of angels and hear how they return your

---

12 *Ṣaḥīḥ al-Bukhārī*, no. 3245.

13 *Ṣaḥīḥ Muslim*, no. 2834.

14 *Musnad Aḥmad*, no. 10524.

15 *Musnad Aḥmad*, no. 7486. The chain contains "ʿIyāḍ b. Dīnār from his father," which is a mistake. See the editor's comments on this route.

16 Ibn al-Bakhtarī, *Majmūʿ fīhi muṣannafāt Abī Jaʿfar b. al-Bakhtarī* (Beirut: Dār al-Bashāʾir al-Islāmiyya, 2001), 229, no. 217. Several scholars have mentioned that ʿUqba b. Abī al-Ḥasnāʾ is an unknown narrator. See al-Dhahabī, *Mīzān al-iʿtidāl*, no. 5685.

17 *Musnad Aḥmad*, no. 10548.

18 Abū Nuʿaym al-Aṣfahānī, *Ṣifat al-janna* (Damascus: Dār al-Maʾmūn, 1995), 2:92, no. 250; al-Bayhaqī, *al-Baʿth wa-l-nushūr* (Riyadh: Maktabat Dār al-Ḥijāz, 2013), no. 437; Shams al-Dīn al-Dhahabī, *Ithbāt al-shafāʿa* (Riyadh: Aḍwāʾ al-Salaf, 2000), 45, no. 34.

19 *Musnad Aḥmad*, no. 11126; *Jāmiʿ al-Tirmidhī*, no. 2535.

20 *Musnad al-Bazzār* (Medina: Maktabat al-ʿUlūm wa-l-Ḥikma, 2009), 5:243, no. 1855.

21 *Ṣaḥīḥ Muslim*, no. 2835; *Ṣaḥīḥ Ibn Ḥibbān*, no. 7435; *Sharḥ al-sunna*, no. 4375. The version of Jābir contains only the middle portion of the hadith, in which the people of Paradise are described.

22 Al-Bayhaqī, *al-Baʿth wa-l-nushūr*, no. 406. Al-Bayhaqī said that this route is unreliable due to the narrators Yazīd al-Raqāshī and Saʿīd b. Zarbī.

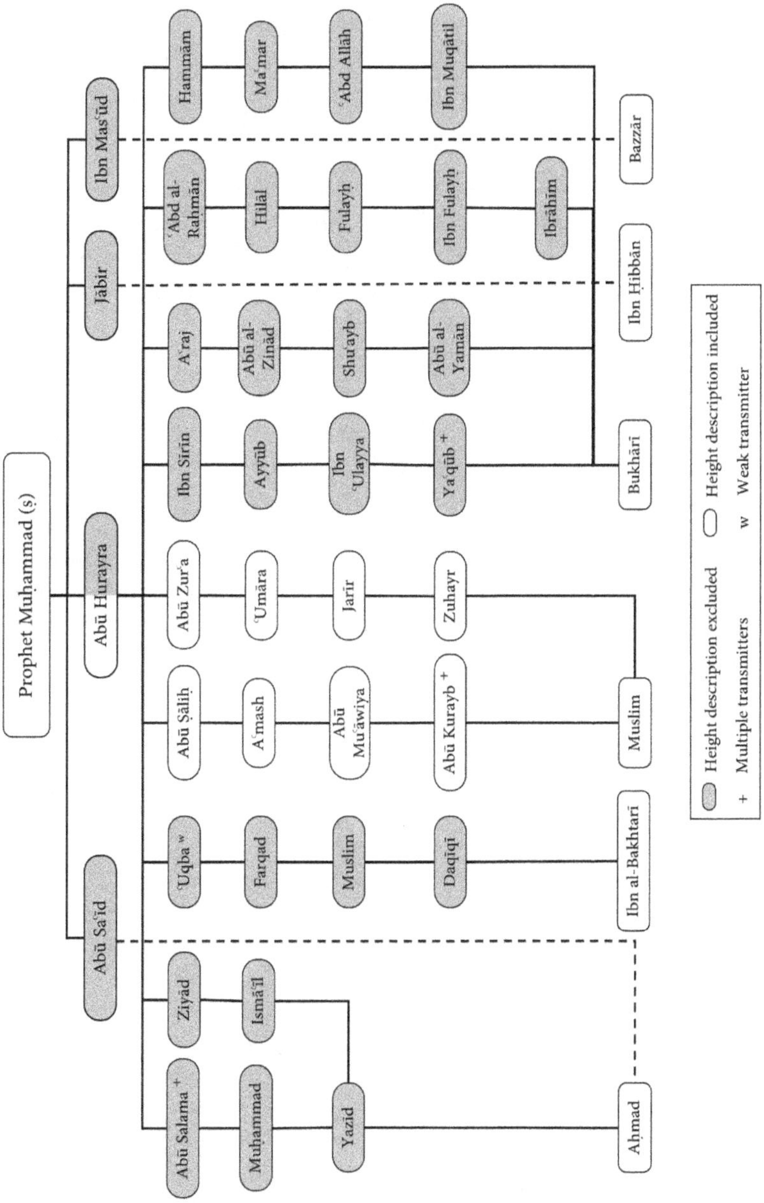

Figure 5: Transmission of Hadith 1

greeting; that is the greeting for you and your progeny." [Adam] said, "Peace be upon you," and they replied, "Peace be upon you and Allah's mercy." Thus, they added "and Allah's mercy." Everyone who enters Paradise will be in Adam's image, and people have been decreasing until this day."[23]

However, al-Shaʿbī, Abū Salama, Abū Ṣāliḥ, Saʿīd al-Maqburī, and Yazīd b. Hurmuz narrate this hadith from Abū Hurayra without the description of sixty cubits or the gradual decrease in height.[24] This version of the hadith begins with Adam sneezing and praising

---

23 *Ṣaḥīḥ al-Bukhārī*, no. 3326; *Ṣaḥīḥ Muslim*, no. 2841; *Jāmiʿ Maʿmar b. Rāshid*, in *Muṣannaf ʿAbd al-Razzāq*, no. 19435; Hammām b. Munabbih, *Ṣaḥīfat Hammām b. Munabbih* (Beirut: al-Maktab al-Islāmī, 1987), no. 58. Ibn Mandah describes this hadith as "established unanimously by the experts of hadith," a phrase that he uses to describe other hadith as well. His grading should be understood in view of the chapter in which the entire hadith was cited. See Ibn Mandah, *al-Radd ʿalā al-Jahmiyya* (Medina: Maktabat al-Ghurabāʾ al-Athariyya, 1994), 42.

24 Al-Nasāʾī narrates these routes via Muḥammad b. Khalaf → Ādam b. Abī Iyās → Abū Khālid Sulaymān b. Ḥayyān; *al-Sunan al-kubrā*, no. 9977. A number of experts have deemed Abū Khālid a reliable transmitter, but others have noted that he was not among the hadith experts and that his additions should not be taken as evidence. See Ibn Ḥajar al-ʿAsqalānī, *Tahdhīb al-Tahdhīb* (Deccan: Maṭbaʿat Dāʾirat al-Maʿārif al-Niẓāmiyya, 1325 AH), 4:181. Abū Khālid is corroborated in some of his routes. Al-Tirmidhī (no. 3368) and al-Nasāʾī (no. 9975) narrate it via Ṣafwān b. ʿĪsā → al-Ḥārith b. Abī Dhubāb → Saʿīd al-Maqburī → Abū Hurayra, and Abū Yaʿlā (no. 6580) narrates it via ʿAmr [*sic*; ʿUmar] b. Muḥammad → Ismāʿīl b. Rāfiʿ → Saʿīd al-Maqburī → Abū Hurayra, without the words referenced. Al-Nasāʾī writes that Ṣafwān's route is erroneous and Muḥammad b. Khalaf's is objectionable; he prefers the non-prophetic version from ʿAbd Allāh b. Salām; *al-Sunan al-kubrā*, no. 9976; cf. al-Dāraquṭnī, *al-ʿIlal*, 8:147. For a study of the various routes of this version of the hadith, see Abū Ḥudhayfa, *Anīs al-sārī fī taḥqīq wa-takhrīj al-aḥādīth allatī dhakarahā al-Ḥāfiẓ Ibn Ḥajar al-ʿAsqalānī fī Fatḥ al-Bārī* (Beirut: Muʾassasat al-Rayyān, 2005), 2:1589–1594; see also the editor's annotations on Ibn Ḥajar, *al-Maṭālib al-ʿāliya*, 14:230–235. Though the route with the inclusion of the two clauses on height is mentioned in the *Ṣaḥīḥayn*, it is worth reading the comments of ʿAbd al-Rashīd al-Nuʿmānī, *al-Imām Ibn Mājah wa-kitābuhu al-Sunan* (Beirut: Dār al-Bashāʾir al-Islāmiyya, 1998), 108–109.

Allah, after which he greets the angels.[25] An unreliable and truncated version is related from ʿAṭāʾ al-Khurāsānī from Abū Hurayra with only the words "Whoever enters Paradise will be in Adam's form. People have since been decreasing till this day."[26] Four other routes are relevant for our purposes here.

(1) It is transmitted via Ibrāhīm b. Saʿīd [al-Jawharī] → Abū Khālid [Sulaymān] → Ibn ʿAjlān [al-Qurashī] → Saʿīd [al-Maqburī] → Abū Hurayra → the Prophet (ṣ), who said, "Allah created Adam; his height was sixty cubits."[27] Ibn ʿAjlān was reliable, but Yaḥyā al-Qaṭṭān said that he was not proficient in the hadith of Saʿīd al-Maqburī; he would confuse the hadith of Saʿīd from his father with those from Abū Hurayra.[28]

(2) A version of the hadith without the height description is transmitted via al-Layth → Ibn ʿAjlān → Saʿīd al-Maqburī → his father → ʿAbd Allāh b. Salām as his own words (*mawqūf*).[29] Al-Nasāʾī regards the non-prophetic version from ʿAbd Allāh b. Salām as the correct version from Saʿīd al-Maqburī.[30]

(3) Mūsā b. Abī ʿUthmān narrates via his father from Abū Hurayra that "God created Adam in His/his image." ʿAbd

---

25 The first part about sneezing is corroborated by Hafṣ b. ʿĀṣim from Abū Hurayra, and it is also narrated from Anas b. Mālik, both as a prophetic hadith and as his own statement. See *Ṣaḥīḥ Ibn Ḥibbān*, nos. 6164, 6164; *al-Mustadrak*, no. 7682; Ibn Abī ʿĀṣim, al-Sunna (Beirut: al-Maktab al-Islāmī, 1400 AH), no. 596.

26 Isḥāq b. Rāhawayh, *Musnad*, 1:390, no. 434. This chain is weak because Kulthūm b. Muḥammad has been impugned and ʿAṭāʾ al-Khurāsānī did not hear from Abū Hurayra. See the editor's comments in al-Ṭabarānī, *Musnad al-Shāmiyyīn* (Beirut: Muʾassasat al-Risāla, 1989), no. 2332, 2369; cf. Abū Ḥudhayfa, *Anīs al-sārī*, 2:1606.

27 *Musnad al-Bazzār*, no. 8511.

28 Al-Bukhārī, *al-Tārīkh al-kabīr* (Hyderabad: Dāʾirat al-Maʿārif al-ʿUthmāniyya, n.d.), no. 603; al-Dhahabī, *Siyar aʿlām al-nubalāʾ*, 6:322.

29 Al-Nasāʾī, *al-Sunan al-kubrā*, no. 9976.

30 Al-Nasāʾī, *al-Sunan al-kubrā*, no. 9976; Aḥmad b. Ḥanbal, *al-ʿIlal*, nos. 5632, 5633.

Allāh, the son of Imam Aḥmad, says that "I found in my father's copy [the added phrase] 'with a height of sixty cubits,' but I do not know if he narrated it to us."[31] Other sources that cite Mūsā's transmission do not include that phrase.[32]

(4) Ja'far al-Firyābī (d. 301 AH) narrates the hadith via its well-known chain from Hammām b. Munabbih without the words "Everyone who enters Paradise will be in Adam's image, and people have been decreasing until this day."[33] These words, however, are soundly transmitted from Hammām in multiple sources. Therefore, it was likely al-Firyābī who truncated the hadith by omitting the passage about gradual human decrease.

It is reported that Ibn 'Umar said that people have been decreasing in size, age, and character since the era of Noah.[34] Although this post-prophetic report mentions the gradual decrease of height, it does not specify a particular height. Figure 6 outlines the routes of Hadith 2, highlighting the inclusion and exclusion of the relevant words.

---

31 *Musnad Aḥmad*, no. 8291.

32 'Abd b. Ḥumayd, *al-Muntakhab min Musnad 'Abd b. Ḥumayd* (Riyadh: Dār Balansiyya, 2002), 2:336, no. 1425.

33 Al-Firyābī, *Kitāb al-Qadar*, 32, no. 3.

34 This report is transmitted via Muḥammad b. Sūqa, al-Ḥakam, and Salama → Mujāhid → Ibn 'Umar. See Nu'aym b. Ḥammād, *al-Fitan* (Cairo: Maktabat al-Tawḥīd, 1412 AH), 2:704; Ibn Abī Ḥātim al-Rāzī, *Tafsīr Ibn Abī Ḥātim al-Rāzī* (Mecca: Maktabat Nizār Muṣṭafā al-Bāz, 1419), 9:3041; Abū Nu'aym al-Aṣfahānī, *Ḥilyat al-awliyā' wa-ṭabaqāt al-aṣfiyā'* (Beirut: Dār al-Kutub al-'Ilmiyya, 1988).

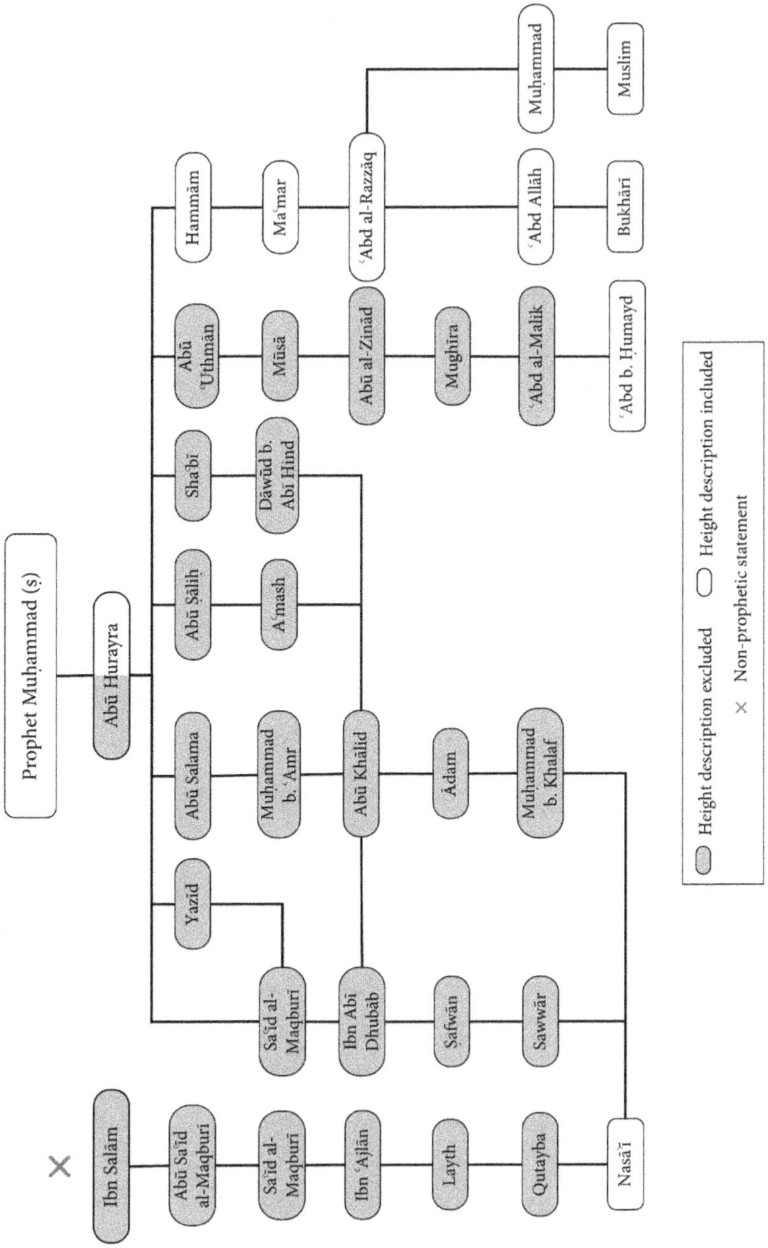

Figure 6: Transmission of Hadith 2

Hadith 3: Physical features in Paradise

The third hadith describes the physical features of those who will enter Paradise. Variations of this hadith are narrated by Abū Hurayra, Anas b. Mālik, Muʿādh b. Jabal, al-Miqdām b. Maʿdīkarib, and al-Miqdād b. al-Aswad, among others. From the Companions, the mention of Adam's height as sixty cubits is recorded in the routes of Abū Hurayra and Anas, but these routes are problematic.

The version of Abū Hurayra is transmitted through the following route: Ḥammād b. Salama → ʿAlī b. Zayd b. Judʿān → Saʿīd b. al-Musayyab → Abū Hurayra → the Prophet (ṣ), who said, "Those destined for Paradise will enter Paradise without body hair or beards, bright, with wavy hair and eyes anointed with kohl, thirty-three years of age, in the image of Adam, sixty cubits [in length] and seven cubits in width."[35] The majority of Ḥammād's students narrate the hadith from him with the description of Adam's height. These students are Yaḥyā b. al-Sakan,[36] Yazīd b. Hārūn,[37] ʿAffān b. Muslim,[38] ʿUbayd Allāh b. Muḥammad,[39] Hudba b. Khālid,[40] and al-Ḥasan b. Sufyān.[41] On the other hand, two hadith experts, Ādam b. Abī Iyās and Abū Salama al-Tabūdhakī, narrate the hadith from Ḥammād without the words

---

35 *Musnad Aḥmad*, no. 7933; *Muṣannaf Ibn Abī Shayba* (Riyadh: Maktabat al-Rushd, 1988), no. 34006. Some routes contain the words "seventy cubits." See *Musnad Aḥmad*, no. 8524, 9375. There is a disagreement among the narrators whether the report is *mursal* or *muttaṣil*. Regardless, Abū Ḥātim al-Rāzī observes that both versions are accurate. See al-Rāzī, *Kitāb al-ʿIlal*, 5:502–503. The supposed route of Muḥammad b. Ziyād referenced to *Musnad Aḥmad* by secondary sources is a typo. The correct name is ʿAlī b. Zayd, as found in the extant manuscripts of the *Musnad* and based on al-Ṭabarānī's negation of this report being transmitted by anyone besides ʿAlī b. Zayd. See Farīd, *Athar al-ʿilm al-tajrībī*, 136.

36 Ibn Saʿd, *al-Ṭabaqāt al-kubrā*, 1:32. Yaḥyā transmits this via Saʿīd b. al-Musayyab as a disconnected hadith.

37 *Muṣannaf Ibn Abī Shayba*, no. 34006.

38 *Musnad Aḥmad*, no. 8524. This route contains the words "seventy cubits."

39 Abū al-Qāsim al-Ṭabarānī, *al-Muʿjam al-ṣaghīr* (Beirut: al-Maktab al-Islāmī, 1985), no. 808; al-Ṭabarānī, *al-Muʿjam al-awsaṭ* (Cairo: Dār al-Ḥaramayn, 1415 AH), no. 5422.

40 Al-Bayhaqī, *al-Baʿth wa-l-nushūr*, no. 997.

41 Al-Aṣfahānī, *Ṣifat al-janna*, 2:99, no. 255.

"sixty cubits [in length] and seven cubits in width."[42] Regardless of its exact wording, the hadith is unreliable due to the weakness of ʿAlī b. Zayd b. Judʿān, as stated by Abū al-Faḍl al-Maqdisī (d. 507 AH).[43] Ibn Ḥibbān said that ʿAlī b. Zayd erred so frequently that he should be abandoned.[44] ʿĀmir al-Aḥwal narrates a shorter version of this hadith via Shahr b. Ḥawshab → Abū Hurayra → the Prophet (ṣ), without any mention of Adam's height.[45] Abū ʿĪsā al-Tirmidhī (d. 279 AH) grades this route as "rare" (*gharīb*).[46]

The route of Muʿādh is transmitted via ʿImrān al-Qaṭṭān → Qatāda → Shahr b. Ḥawshab → ʿAbd al-Raḥmān b. Ghanm → Muʿādh b. Jabal → Prophet (ṣ), who said, "Those destined for Paradise will enter Paradise without body hair or beards, eyes anointed with kohl, and thirty or thirty-three years of age."[47] There is no mention of Adam's height in this route. Al-Tirmidhī grades the hadith as "fair, rare" (*ḥasan gharīb*) and mentions that some of Qatāda's students narrate

42 Al-Rāzī, *Kitāb al-ʿIlal*, 5:502–503. Although al-Rāzī does not mention the wording of Abū Salama's transmission, it can be assumed that the wording is the same as the version of Ādam b. Abī Iyās, because al-Rāzī combines the two routes and only highlights the distinctions in the chain. Moreover, Abū Ḥātim's statement that both the connected and the disconnected versions are sound (*ṣaḥīḥayn*) should not be understood as an authentication of the chain; rather, he is saying that the hadith has been accurately transmitted via connected and disconnected chains, regardless of its grading.

43 Abū al-Faḍl al-Maqdisī, *Dhakhīrat al-ḥuffāẓ* (Riyadh: Dār al-Salaf, 1996), 4:2020. On the status of Ḥammād b. Salama, see Muḥammad Zāhid al-Kawtharī, "Introduction," in *Kitāb al-Asmāʾ wa-l-ṣifāt* (Cairo: al-Maktaba al-Azhariyya, n.d.), 3.

44 For Ibn Ḥibbān's criticism alongside other negative remarks from hadith experts, see Ibn al-Jawzī, *al-Mawḍūʿāt* (Medina: al-Maktaba al-Salafiyya, 1968), 2:26; Ibn al-Jawzī, *al-Ḍuʿafāʾ wa-l-matrūkūn*, 2:193, no. 2373. For a counter perspective on ʿAlī b. Zayd, see Ḥātim al-ʿAwnī, *al-Mursal al-khafī* (Riyadh: Dār al-Hijra, 1997), 1:306–322.

45 ʿAbd Allāh al-Dārimī, *Sunan al-Dārimī* (Riyadh: Dār al-Mughnī, 2000), no. 2868.

46 *Jāmiʿ al-Tirmidhī*, no. 2539.

47 *Musnad Aḥmad*, no. 22106. Al-Haythamī states that the hadith cited in *Musnad Aḥmad* is fair and contains a connected chain. See Nūr al-Dīn al-Haythamī, *Majmaʿ al-zawāʾid* (Cairo: Maktabat al-Qudsī, 1994), 10:398.

it from him via a disconnected chain.[48] Figure 7 outlines the routes of Hadith 3.1 via Abū Hurayra and Muʿādh, highlighting the inclusion and exclusion of the relevant words.

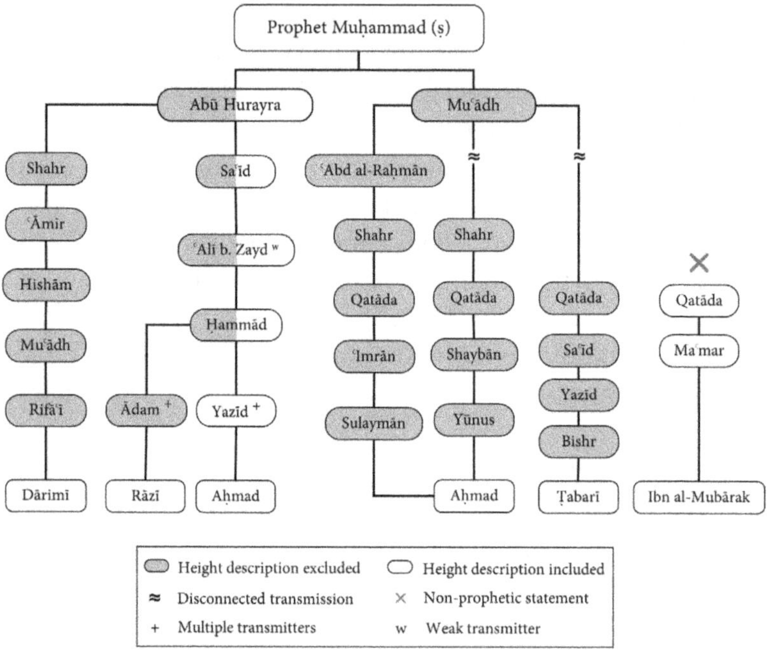

Figure 7: Transmission of Hadith 3.1

A similar hadith has been transmitted on the authority of Anas. Ibn Abī al-Dunyā narrates via Rawwād b. al-Jarrāḥ → al-Awzāʿī → Hārūn b. Riʾāb → Anas → the Prophet (ṣ), who said, "People will enter Paradise with the height of Adam, sixty cubits per the royal cubit (*dhirāʿ al-malik*) . . ."[49] Three other students of al-Awzāʿī narrate the

48 *Jāmiʿ al-Tirmidhī*, no. 2545. For the same hadith with a disconnected chain from Qatāda, see *Musnad Aḥmad*, no. 22024; al-Ṭabarī, *Jāmiʿ al-Bayān*, 20:260. Al-Zaylaʿī records al-Tirmidhī's comments as only "rare." However, al-Mizzī records his comments as "fair, rare." See ʿAbd Allāh b. Yūsuf al-Zaylaʿī, *Takhrīj aḥādīth al-Kashshāf* (Riyadh: Dār Ibn Khuzayma, 1414 AH), 3:408; Jamāl al-Dīn al-Mizzī, *Tuḥfat al-ashrāf bi-maʿrifat al-aṭrāf* (Beirut: Dār al-Gharb al-Islāmī, 1999), 8:94, no. 11336.

49 Ibn Abī al-Dunyā, *Ṣifat al-janna* (Beirut: Muʾassasat al-Risāla, 1997), 163, no. 218.

hadith from him without the words "with the height of Adam, sixty cubits per the royal cubit":

(1) al-Ṭabarānī (d. 360 AH) narrates from ʿUmar b. ʿAbd al-Wāḥid → al-Awzāʿī → Hārūn b. Riʾāb → Anas → the Prophet (ṣ) without those words;[50]

(2) al-Bukhārī narrates from al-Walīd b. Muslim from al-Awzāʿī via the same chain, without those words;[51] and

(3) Abū al-Qāsim Tammām (d. 414 AH) narrates it from Naṣr b. al-Ḥajjāj from al-Awzāʿī in the same manner.[52]

The version of this hadith narrated by the majority of al-Awzāʿī's students without the mention of sixty cubits is more reliable. Ibn Abī al-Dunyā's chain contains Rawwād b. al-Jarrāḥ, an impugned transmitter,[53] and al-Ṭabarānī's chain was graded as reliable.[54] There is disagreement over whether Hārūn heard from Anas.[55] Abū Nuʿaym al-Aṣfahānī (d. 430 AH) mentions that some routes add an unnamed link between Hārūn and Anas.[56]

Abū Nuʿaym narrates via Ayyūb al-Wazzān → Ghassān b. ʿUbayd → Abū ʿĀtika → Anas → the Prophet (ṣ) a similar hadith without the height description.[57] However, Abū ʿĀtika's weakness as a narrator "is a point of consensus."[58] Abū al-Ḥasan Baḥshal (d. 292 AH) narrates via ʿIkrima → Yaḥyā b. Abī Kathīr → Anas → the Prophet (ṣ),

---

50 Al-Ṭabarānī then says, "No one besides ʿUmar b. ʿAbd al-Wāḥid narrated from al-Awzāʿī, and Maḥmūd b. Khālid is alone in narrating it." See al-Ṭabarānī, *al-Muʿjam al-ṣaghīr*, 2:278, no. 1164.

51 Al-Bukhārī, *al-Tārīkh al-kabīr*, 8:219.

52 Abū al-Qāsim Tammām, *al-Fawāʾid* (Riyadh: Maktabat al-Rushd, 1992), 347, no. 891.

53 Ibn ʿAdī, *al-Kāmil*, 4:114–120, no. 684.

54 Al-Haythamī, *Majmaʿ al-zawāʾid*, 10:399.

55 Jamāl al-Dīn al-Mizzī, *Tahdhīb al-kamāl fī asmāʾ al-rijāl* (Beirut: Muʾassasat al-Risāla, 1980), 30:82; see also the editor's annotations on Abū Ṭāhir al-Silafī, *al-Ṭuyūriyyāt* (Riyadh: Maktabat Aḍwāʾ al-Salaf, 2004), no. 1329.

56 He writes, "Others have narrated via al-Awzāʿī from Hārūn, who said, 'It was narrated to me from someone who heard Anas.'" See al-Aṣfahānī, *Ḥilyat al-awliyāʾ*, 3:56.

57 Al-Aṣfahānī, *Ṣifat al-janna*, 2:108, no. 262.

58 Al-Dhahabī, *Mīzān al-iʿtidāl*, 2:335, 4:542.

who said, "The inhabitants of Paradise will be thirty-three years old, fair, beardless, eyes anointed with kohl, and sixty cubits tall." Though sixty cubits is mentioned in this route, it is unrelated to the height of Adam.[59]

The routes of al-Miqdām and al-Miqdād are as follows: Yazīd b. Sinān → Abū Yaḥyā Salīm al-Kalāʿī → al-Miqdām b. Maʿdīkarib → the Prophet (ṣ) with the words "in the form of Adam" without any mention of his height.[60] After narrating this hadith, Yaʿqūb al-Fasawī (d. 277 AH) makes it a point to distinguish al-Miqdām from the more famous Companion al-Miqdād b. al-Aswad. However, al-Ṭabarānī narrates the hadith via Yazīd → al-Kalāʿī → al-Miqdād b. al-Aswad → the Prophet (ṣ),[61] and then narrates it from al-Miqdām in a separate chapter.[62] The common link for both routes is Yazīd b. Sinān, who is described as "weak, with slight accreditation."[63] Al-Ṭabarānī narrates a similar hadith, which has been graded as fair,[64] through a different route via ʿAbd Allāh b. Sālim → al-Zubaydī → al-Kalāʿī → al-Miqdām → the Prophet (ṣ) with the words "in the appearance (*masḥat*) of Adam" but no mention of his height.[65] Figure 8 outlines the routes

---

59 Abū al-Ḥasan Baḥshal, *Tārīkh Wāsiṭ* (Beirut: ʿĀlam al-Kutub, 1406 AH), 210–211; Abū Bishr al-Dūlābī, *al-Kunā wa-l-asmāʾ* (Beirut: Dar al-Kutub al-ʿIlmiyya, 1999), 1:285, no. 1011.

60 Yaʿqūb al-Fasawī, *al-Maʿrifa wa-l-tārīkh* (Beirut: Muʾassasat al-Risāla, 1981), 2:160–161. Ibn Ḥajar relates that Abū Yaʿlā also narrates al-Miqdām's hadith via ʿAlī b. Mushir from Yazīd b. Sinān. See Ibn Ḥajar, *al-Maṭālib al-ʿāliya,* 18:724.

61 Al-Ṭabarānī, *al-Muʿjam al-kabīr,* 20:256. The name al-Miqdād in the chain is not a typographical error since al-Ṭabarānī narrates the hadith in a section dedicated to al-Miqdād. Salīm b. ʿĀmir is a reliable narrator who heard from both al-Miqdām and al-Miqdād. See al-Nawawī, *Tahdhīb al-asmāʾ wa-l-lughāt* (Beirut: Dār al-Kutub al-ʿIlmiyya, n.d.), 1:232. That said, if a confusion did occur, it would have been caused by one of the narrators in the chain. Ibn Rajab al-Ḥanbalī notes that the people of the Levant would call al-Miqdām b. Maʿdīkarib by the name "al-Miqdād," and when no further details were given, he would be confused for al-Miqdād b. al-Aswad. This resulted in several of their hadith being conflated. See Ibn Rajab, *Fatḥ al-Bārī,* 4:52.

62 Al-Ṭabarānī, *al-Muʿjam al-kabīr,* 20:280.

63 Al-Haythamī, *Majmaʿ al-zawāʾid,* 10:334.

64 Al-Haythamī, *Majmaʿ al-zawāʾid,* 10:331.

65 Al-Ṭabarānī, *al-Muʿjam al-kabīr,* 20:280.

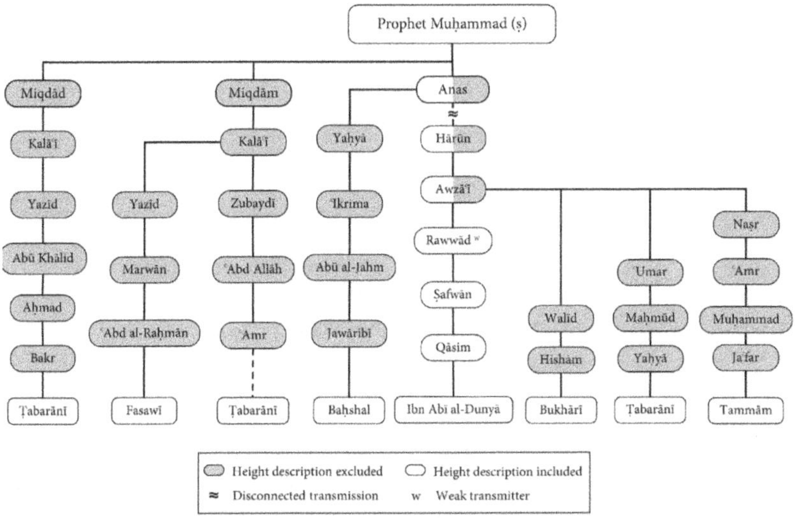

Figure 8: Transmission of Hadith 3.2

There are a few more hadith that are relevant to our discussion. The first group of hadith are transmitted from several Companions without the height description:

(1) Ibn Mandah (d. 395 AH) narrates that Abū Lubāba, Ibn Busr, and Wāthila b. al-Asqaʿ said that the Prophet (ṣ) delivered a sermon about the afterlife in which he mentioned that "its inhabitants will not die; they will be thirty-three years old, beardless, and their eyes anointed with kohl."[66] There is no mention of Adam's height in this narration.

(2) Abū Nuʿaym narrates the hadith via Jaʿfar → his father → Abū Rajā' → Ibn ʿAbbās → the Prophet (ṣ) without the height description.[67] However, both Jaʿfar b. Jasr and his father Jasr b. Farqad are unreliable.[68]

66 Ibn Mandah, *Min Amālī Ibn Mandah*, fol. 20v., MS al-Assad National Library, no. 3772.

67 Al-Aṣfahānī, *Ṣifat al-janna*, 2:106, no. 260.

68 Al-Dhahabī, *Mīzān al-iʿtidāl*, 1:398–399, 403–404.

(3) Abū al-Haytham narrates from Abū Sa'īd al-Khudrī from the Prophet (ṣ) that the young and the old who are destined for Paradise will "be thirty-three years old and will never grow older," without any mention of the other features.[69]

(4) It is related from Ibn Mas'ūd,[70] Ibn 'Abbās,[71] and Jābir that the Prophet (ṣ) said, "People will be raised on the Day of Judgment without body hair or beards, thirty-three years old, except for Moses son of Amram, for he will have a beard until his navel."[72] There is no mention of the height description here, but every route of this hadith is riddled with errors and some routes have been described as outright fabrications.[73]

There are two reports that include the height description. Al-Ḥasan al-Baṣrī narrates the hadith via a disconnected chain from the Prophet (ṣ) with the description of Adam's height.[74] Ma'mar

---

69 Ibn al-Mubārak, *Kitāb al-Raqā'iq*, 2:771, no. 1659; *Jāmi' al-Tirmidhī*, no. 2562 with the comment that "this is a rare hadith; I do not recognize it except from the hadith of Rishdīn."

70 Ibn Ḥajar mentions that Ibn Mas'ūd's route is transmitted by al-Ṭabarānī via a weak chain. See Ibn Ḥajar al-'Asqalānī, *Masā'il ajāba 'anhā al-Ḥāfiẓ Ibn Ḥajar al-'Asqalānī* (Cairo: Dār al-Imām Aḥmad, 2007), 25–27. I could not locate Ibn Mas'ūd's route in al-Ṭabarānī's extant works or in al-Haythamī's *Majma' al-zawā'id*. Contrary to multiple secondary sources, Burhān al-Dīn al-Nājī (d. 900 AH) quotes Ibn Ḥajar as referencing al-Ṭabarī instead of al-Ṭabarānī; al-Nājī, *Ḥuṣūl al-bughya li-l-sā'il hal li-aḥad fī al-janna liḥya* (Beirut: Dār al-Bashā'ir al-Islāmiyya, 2004), 23, 28.

71 Abū Nu'aym al-Aṣfahānī narrates the route of Ibn 'Abbās via Mujāshi' b. 'Amr, an extremely unreliable narrator. See al-Aṣfahānī, *Ṣifat al-janna*, 2:108; al-Dhahabī, *Mīzān al-i'tidāl*, 3:436. Abū al-Maḥāsin al-Rūyānī (d. 502 AH) records this route via a different chain, but it is riddled with flaws. See al-Nājī, *Ḥuṣūl al-bughya*, 22–23.

72 The route of Jābir has been described as a baseless fabrication. See Ibn Ḥibbān, *Kitāb al-Majrūḥīn* (Aleppo: Dār al-Wa'y, 1396 AH), no. 483.

73 Ibn al-Jawzī, *al-Mawḍū'āt*, 3:257–259; cf. Ibn 'Irāq, *Tanzīh al-sharī'a al-marfū'a 'an al-akhbār al-shanī'a al-mawḍū'a* (Beirut: Dār al-Kutub al-'Ilmiyya, 1399 AH), 2:384.

74 Yaḥyā b. Sallām, *Tafsīr*, 2:815. It is in this route that al-Ḥasan al-Baṣrī suspended judgment on the exact height of Adam. Regarding Yaḥyā b. Sallām, al-Dāraquṭnī

narrates that Qatāda said, "The inhabitants of Paradise will be thirty years old . . . in the form of Adam, and his height was sixty cubits."[75] This report, however, is a post-prophetic statement of Qatāda.[76]

Hadith 4: Like an enormous date palm

The fourth hadith describes Adam as an enormous date palm. Variations of this hadith are narrated by Ubayy b. Kaʿb, Anas b. Mālik, and ʿAbd Allāh b. ʿAmr. Though certain routes of Ubayy's hadith include the sixty cubits description, the hadith has been criticized as is the case with the transmission of this hadith from the other Companions.

The primary source for this hadith is Ubayy. Ibn Jarīr al-Ṭabarī (d. 310 AH) transmits it via al-Ḥasan al-Baṣrī → Ubayy b. Kaʿb → the Prophet (ṣ), who said, "Adam was a tall man like an enormous date palm (*nakhla saḥūq*), sixty cubits, with plenty of hair."[77] However, authoritative transmitters from al-Ḥasan, such as Qatāda b. Diʿāma, narrate this report without the words "sixty cubits."[78] Even the version

---

mentions that he is weak, while Abū Zurʿa said that there is nothing wrong with him though he occasionally makes mistakes. See al-Dāraquṭnī, *al-Sunan*, nos. 1241, 2283; Abū Zurʿa, *Kitāb al-Ḍuʿafāʾ* (Medina: al-Majlis al-ʿIlmī, 1982), 2:339.

75 *Jāmiʿ Maʿmar b. Rāshid*, no. 29872; Ibn al-Mubārak, *Kitāb al-Raqāʾiq*, 2:771, no. 1660. Ibn al-Mubārak also records al-Zuhrī's statement "It reached us that the people of Paradise will have the build of Adam, and he was sixty cubits." See Ibn al-Mubārak, no. 1661.

76 In the *Jāmiʿ* of Maʿmar, the words "*ʿan Qatāda yarwīhi*" would suggest that Qatāda is attributing the statement to the Prophet (ṣ). However, Ibn al-Mubārak transmits it from Maʿmar from Qatāda as his own words. Even if it is ceded that Qatāda is attributing the words to the Prophet (ṣ), there are at least two links missing in the chain. As noted earlier, Qatāda narrates the hadith from Muʿādh via connected and disconnected chains without the mention of Adam's height. On the usage of "*yarwīhi*" in hadith transmission, see al-Suyūṭī, *Tadrīb al-rāwī*, 3:117–119.

77 Al-Ṭabarī, *Tārīkh al-rusul wa-l-mulūk*, 1:160.

78 Only two routes of this report mention "sixty cubits" as a prophetic hadith. Al-Ṭabarī narrates it via al-Ḥasan b. Dhakwān → al-Ḥasan al-Baṣrī → Ubayy b. Kaʿb → the Prophet (ṣ), as cited above. Abū al-Shaykh narrates it via Muḥammad b. Maymūn → al-Ḥasan → Ubayy → the Prophet (ṣ). See Abū al-Shaykh al-Aṣfahānī, *al-ʿAẓama* (Riyadh: Dār al-ʿĀṣima, 1987), 5:1556. Al-Marwazī narrates it via Muḥammad b. Dhakwān → al-Ḥasan as the words of Ubayy b. Kaʿb. See

of the hadith without the mention of sixty cubits was described by ʿAmr al-Fallās (d. 249 AH) as a detestable hadith (*ḥadīth munkar*).[79] Ibn Kathīr (d. 774 AH) states that it is more correctly transmitted as the words of Ubayy b. Kaʿb (not a prophetic hadith) and that there is a discontinuity in the chain between al-Ḥasan and Ubayy,[80] so it cannot be admitted as evidence on the subject.

The route of Anas is transmitted by Ibn ʿAsākir (d. 571 AH) via Ādam b. Abī Iyās → Shaybān → Qatāda → Anas → the Prophet (ṣ), but it simply describes Adam as "extremely tall" (*ṭuwāl saḥūq*) without the words "sixty cubits" or "date palm."[81] No other route of this hadith comes from Anas. In fact, a similar route is transmitted by Aḥmad via Yūnus al-Mu'addib → Shaybān → Qatāda → al-Ḥasan → Ubayy (instead of Anas), which corresponds to the other routes of this hadith.[82]

As for the route of ʿAbd Allāh b. ʿAmr, Ibn Abī al-Dunyā narrates via Ṣafwān b. ʿAmr → Shurayḥ → Kathīr b. Murra → ʿAbd Allāh b. ʿAmr a lengthy post-prophetic report in which Adam is likened to an enormous date palm without any mention of sixty cubits.[83] However, the chain of transmission for this report is extremely problematic.[84]

Muḥammad b. Naṣr al-Marwazī, *Taʿẓīm qadr al-ṣalāh* (Medina: Maktabat al-Dār, 1986), 2:842, no. 852. More reliable transmitters from al-Ḥasan, like Qatāda, do not mention "sixty cubits." Some routes mention it as a prophetic hadith, while others, like al-ʿAṭṭār via al-Ḥasan, have it as the words of Ubayy. See Yaḥyā b. Sallām, *Tafsīr*, 1:285; Aḥmad b. Ḥanbal, *al-Zuhd* (Beirut: Dār al-Kutub al-ʿIlmiyya, 1999), no. 265; al-Ṭabarī, *Jāmiʿ al-bayān*, 12:354; al-Rāzī, *Tafsīr Ibn Abī Ḥātim*, no. 388. For the route of al-ʿAṭṭār, see Ibn Saʿd, *al-Ṭabaqāt al-kubrā*, 1:28, 31–32.

79 Ibn ʿAdī, *al-Kāmil*, 1:547; cf. al-Dhahabī, *Mīzān al-iʿtidāl*, 1:191.

80 Ibn Kathīr, *Tafsīr*, 3:397, 5:321. For a detailed analysis of this report, see Abū Ḥudhayfa, *Anīs al-sārī*, 2:1585–1588. On the problem with Ibn Ḥajar's grading of a particular route of this hadith as fair, see Abū Isḥāq al-Ḥuwaynī, *Tanbīh al-hājid ilā mā waqaʿa min al-naẓar fī kutub al-amājid* (Beirut: al-Maḥajja, n.d.), 1:453–454.

81 Ibn ʿAsākir, *Tārīkh Dimashq* (Beirut: Dār al-Fikr, 1995), 7:404; Ibn Kathīr, *al-Bidāya wa-l-nihāya*, 1:78.

82 Aḥmad, *al-Zuhd*, no. 265.

83 Ibn Abī al-Dunyā, *Ḥusn al-ẓann* (Riyadh: Dār Ṭayba, 1988), no. 80.

84 Shams al-Dīn al-Sakhāwī, *al-Qawl al-badīʿ fī al-ṣalāh ʿalā al-ḥabīb al-shafīʿ* (Beirut: Mu'assasat al-Rayyān, 2002), 263–264.

The narrator ʿAbd Allāh b. Wāqid al-Ḥarrānī is abandoned (*matrūk*) in hadith.[85] Figure 9 outlines the routes of Hadith 4, highlighting the inclusion and exclusion of the relevant words.

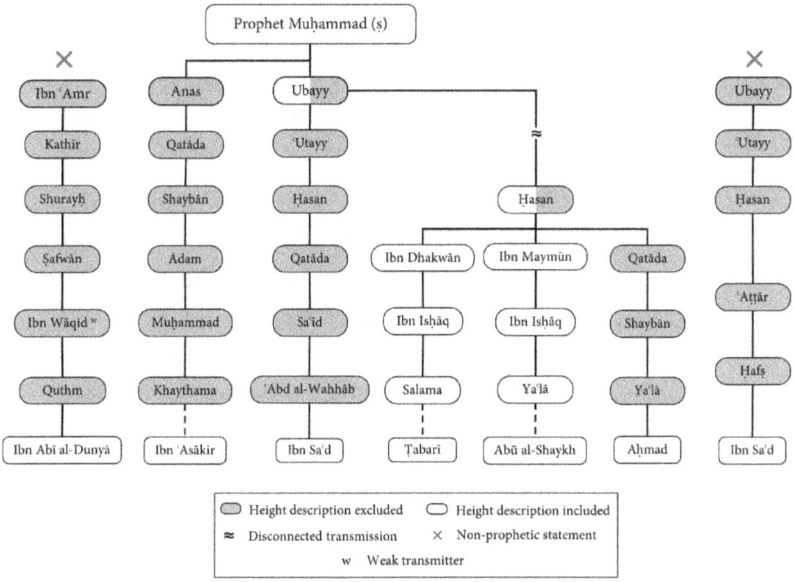

Figure 9: Transmission of Hadith 4

One can argue that the exclusion of the sixty cubits description is irrelevant because the analogy of an "enormous date palm" in itself suggests that Adam was exceptionally tall since date palms average about twenty-three meters.[86] Putting aside the question of the hadith's reliability, such a description appears to be a figure of speech, not a literal measurement of height. The famed philologist al-Aṣmaʿī (d. 216 AH) relates an anecdote about a nomad with impressive poetic skills who was brought to the emir of Kufa. As he entered through the door, al-Aṣmaʿī noted that he was "like an enormous date palm,"

85 Muslim b. al-Ḥajjāj, *al-Kunā wa-l-asmāʾ* (Medina: al-Jāmiʿa al-Islāmiyya, 1984), no. 2805.

86 See https://www.britannica.com/plant/date-palm (accessed 5/12/2021).

which was clearly a metaphor describing his remarkable height.[87] It is also worth remembering that al-Muṭahhar al-Maqdisī interprets "*saḥūq*" as anything above average human height.[88]

Miscellanea

There are three alternatives to the sixty cubits description: twelve, eighteen, and seventy cubits, but none of these reports are reliable:

- ʿAffān narrates via Ḥammād b. Salama → ʿAlī b. Zayd → Saʿīd b. al-Musayyab → Abū Hurayra → the Prophet (ṣ) that Adam was seventy cubits tall and seven cubits in width.[89] As noted earlier, ʿAlī b. Zayd is unreliable.[90] Abū Maymūn al-Ṣūrī narrates via Abū al-Zinād → al-Aʿraj → Abū Hurayra → the Prophet (ṣ), who said, "Allah created Adam in his image, and his height was seventy cubits."[91] Abū Maymūn was considered a liar.[92]

- Ḥammād b. Salama narrates via Saʿīd al-Jurayrī → Abū Naḍra → Shutayr [*sic*; Sumayr] b. Nahār → Abū Hurayra → the Prophet (ṣ), who said, "The poor will enter Paradise before the wealthy . . . in the form of Adam, eighteen cubits tall and seven

87 Abū al-Qāsim al-Naysābūrī, *ʿUqalāʾ al-majānīn wa-l-muwaswasīn* (Beirut: Dār al-Nafāʾis, 1987), 273; see also Najm al-Dīn al-Ṭūfī, *al-Iksīr fī ʿilm al-tafsīr* (Beirut: Dār al-Awzāʿī, 1989), 169.

88 Al-Maqdisī, *Kitāb al-Badʾ wa-l-tārīkh*, 3:22.

89 *Musnad Aḥmad*, nos. 8524, 9374.

90 This report was cited earlier under Hadith 3.

91 Al-Ṭabarānī, *Musnad al-Shāmiyyīn*, no. 3357; Ibn ʿAsākir, *Tārīkh Dimashq*, 10:118. Al-Bayhaqī ascribes the seventy cubits description to *Ṣaḥīḥ Muslim* via Jarīr → ʿUmāra → Abū Zurʿa → Abū Hurayra → the Prophet (ṣ); *al-Baʿth wa-l-nushūr*, no. 907. However, the hadith in *Ṣaḥīḥ Muslim* only mentions sixty cubits (no. 2834). It is transmitted from Ibn ʿAbbās, ʿAbd Allāh b. ʿAmr, and others that Adam's height initially reached the heavens but was then decreased to seventy cubits. Not only are these reports transmitted via extremely unreliable chains, they are also based on a passage from the Talmud, as noted earlier. See al-Ṭabarānī, *al-Muʿjam al-kabīr*, no. 14158; al-Haythamī, *Majmaʿ al-zawāʾid*, no. 5726; Abū Ismāʿīl al-Harawī, *al-Arbaʿīn fī dalāʾil al-tawḥīd* (Medina: n.p., 1984), no. 21.

92 Nāyif al-Manṣūrī, *Irshād al-qāṣī wa-l-dānī ilā tarājim shuyūkh al-Ṭabarānī* (Sharjah: Maktabat Ibn Taymiyya, 2006), 292.

cubits in width."[93] This hadith has been narrated through various routes without any mention of Adam's description.[94] In fact, Shuʿba narrates the hadith from al-Jurayrī via the same chain without the description.[95] In addition, al-Jurayrī became senile in the latter part of his life,[96] and Sumayr b. Nahār was an unknown narrator.[97]

- ʿAdī b. al-Faḍl, a *matrūk* narrator, relates from al-Jurayrī → Abū Naḍra → ʿAqīl b. Sumayr → Abū Hurayra → the Prophet (ṣ) that the poor will enter Paradise before the wealthy in the form of Adam, twelve cubits tall and six cubits in width.[98] Ismāʿīl b. Ibrāhīm narrates this report from al-Jurayrī → Abū Naḍra →

93 Al-Bayhaqī, *al-Baʿth wa-l-nushūr*, nos. 984, 985.

94 See Abū Ḥudhayfa, *Anīs al-sārī*, no. 4706.

95 *Musnad Aḥmad*, no. 10730.

96 Salāḥ al-Dīn al-ʿAlāʾī, *al-Mukhtaliṭīn* (Cairo: Maktabat al-Khānjī, 1996), no. 16.

97 Aḥmad b. Ḥanbal, *al-ʿIlal*, no. 983. There is a disagreement regarding the name of Sumayr b. Nahār, with some saying that his name is Shutayr while others say it is Shumayr. According to al-Bukhārī, the correct name is Sumayr. See al-Bukhārī, *al-Tārīkh al-kabīr*, 2490.

98 Al-Ṭabarānī, *al-Muʿjam al-awsaṭ*, no. 8865; al-Haythamī, *Majmaʿ al-zawāʾid*, 10:260, no. 17894; Ḥasan al-Wāʾilī, *Nuzhat al-albāb fī qawl al-Tirmidhī «wa-fī al-bāb»* (Jeddah: Dār Ibn al-Jawzī, 2005), 5:3194. One wonders if ʿAqīl b. Sumayr in this chain is meant to be Sumayr b. Nahār. Biographers mention that al-Jurayrī also narrated from ʿAqīl b. Sumayr, but they state that ʿAqīl narrated from Ibn ʿUmar and that al-Juryrī's transmission from him was via Sayyār b. Salāma. Furthermore, given ʿAdī b. al-Faḍl's unreliability, al-Jurayrī's senility, and the fact that the other routes of the hadith pass through Sumayr b. Nahār, it is possible that Sumayr was confused for ʿAqīl. On ʿAqīl b. Sumayr, see Ibn Abī Ḥātim al-Rāzī, *Kitāb al-Jarḥ wa-l-taʿdīl* (Deccan: Dāʾirat al-Maʿārif al-ʿUthmāniyya, 1952), 6:218; al-Dāraquṭnī, *al-Muʾtalif wa-l-mukhtalif* (Beirut: Dār al-Gharb al-Islāmī, 1986), 3:1576. It is worth noting that al-Dāraquṭnī mentions ʿAqīl among those with the name "ʿAqīl," who are followed by those with the name "ʿUqayl," which shows the correct vowelization of his name.

an unnamed man → Abū Hurayra as his own words,[99] which al-Dāraquṭnī regards as the accurate version.[100]

The height of Adam as sixty cubits is also mentioned in another hadith regarding the state of people during their reckoning in the afterlife. There are prophetic and post-prophetic versions of this hadith:

(1) al-Tirmidhī narrates via Isrāʾīl b. Yūnus → Ismāʿīl al-Suddī → his father → Abū Hurayra → the Prophet (ṣ) that on the Day of Judgment, both believers and disbelievers "will be enlarged to sixty cubits in the image of Adam" and made to wear crowns indicating their standing before God.[101] Abū Ḥātim al-Rāzī states that this report is more accurately transmitted as the words of Abū Hurayra, not as a prophetic hadith.[102]

(2) Yaḥyā b. Sallām (d. 200 AH) narrates via an unidentified source → Abān b. Abī ʿAyyāsh → Abū al-ʿĀliya → Ubayy b. Kaʿb, who said that on the Day of Judgment the believer will be dressed in luxurious garments, made to wear a crown, and "will be enlarged to sixty cubits, which is the height of Adam," whereas the disbeliever will be enlarged to forty cubits and suffer from torments.[103] Not only is this a

---

99 Ibn al-Mubārak, *Kitāb al-Raqāʾiq*, no. 1476. Ibn al-Mubārak narrates this via Ismāʿīl b. Ibrāhīm → al-Jurayrī → Abū Naḍra → an unnamed man → Abū Hurayra.

100 Al-Dāraquṭnī, *al-ʿIlal al-wārida fī al-aḥādīth al-nabawiyya* (Riyadh: Dār Ṭayba, 1985), no. 2089. Although the wording of the hadith that al-Dāraquṭnī mentions is slightly different, it comprises Adam's description as twelve cubits tall and six cubits in width, and its chain of transmission is the same.

101 *Jāmiʿ al-Tirmidhī*, no. 3136 with the grading "fair, rare."

102 When asked about this hadith, Abū Ḥātim al-Rāzī said, "Isrāʾīl narrates this report as a prophetic hadith whereas al-Thawrī does not attribute it to the Prophet; al-Thawrī had better memory (*aḥfaẓ*)." See al-Rāzī, *Kitāb al-ʿIlal*, no. 1762. Furthermore, al-Suddī's father, ʿAbd al-Raḥmān b. Abī Karīma, is unknown (*majhūl*). See al-Dhahabī, *Mīzān al-iʿtidāl*, no. 4947.

103 Yaḥyā b. Sallām, *Tafsīr*, 1:121–122.

post-prophetic pronouncement, but its chain also contains an unidentified narrator and another who is abandoned.[104]

There are indications of the physical description of Adam in the Prophet's encounter with him during the night journey and ascension (*al-isrāʾ wa-l-miʿrāj*). Al-Ṭabarī relates one version via Abū Jaʿfar al-Rāzī → al-Rabīʿ → Abū al-ʿĀliya → Abū Hurayra → the Prophet (ṣ), who said regarding Adam, "There was a man whose stature was complete (*tāmm al-khalq*) and that was not diminished like the stature of other people."[105] The lengthy hadith in which the above description was mentioned has been criticized for its transmission because of Abū Jaʿfar al-Rāzī and for its objectionable content.[106] In addition to these problems, there is no mention of a specific height in this hadith. The description of "complete stature" seems to imply that Adam did not go through the physical changes that is normal for other humans.

Results and preliminary thoughts

Let us now reexamine the point of conflict. On the one hand, there are authentic solitary (*āḥād*) hadith that suggest (1) that Adam was sixty cubits tall and (2) that humankind then gradually decreased in height. On the other hand, there are archaeological and scientific concerns with accepting these two clauses at face value. Based on the preceding analysis, we learn that the majority of transmitters narrate the various hadith on the subject without the contentious passages regarding height. The dominant position among legal theorists is that it is valid to prioritize one of two conflicting sets of reliable reports based on the preponderance of its narrators.[107] As Abū al-Muẓaffar

104 On the status of Abān b. Abī ʿAyyāsh, see al-Dhahabī, *Mīzān al-iʿtidāl*, 1:10; al-Mizzī, *Tahdhīb al-kamāl*, no. 142.

105 Al-Ṭabarī, *Jāmiʿ al-bayān*, 17:337.

106 Ibn Kathīr, *Tafsīr*, 5:36.

107 Ghāzī al-ʿUtaybī, "al-Tarjīḥ bi-kathrat al-ruwāh: dirāsa uṣūliyya taṭbīqiyya," *Majallat Jāmiʿat Umm al-Qurā li-ʿUlūm al-Sharīʿa wa-l-Dirāsāt al-Islāmiyya* 44 (1429 AH): 311–322; Ṣāliḥ al-Nahhām, *al-Ikhtilāf al-uṣūlī fī al-tarjīḥ bi-kathrat al-adilla wa-l-ruwāh wa-atharuhu* (Doha: Wizārat al-Awqāf wa-l-Shuʾūn al-Islāmiyya, 2011), 318–321.

al-Samʿānī (d. 489 AH) contends, the preponderance of narrators adds strength to the reliability of a hadith because "the transmission of the majority is epistemically stronger and further from error than the transmission of the minority."[108]

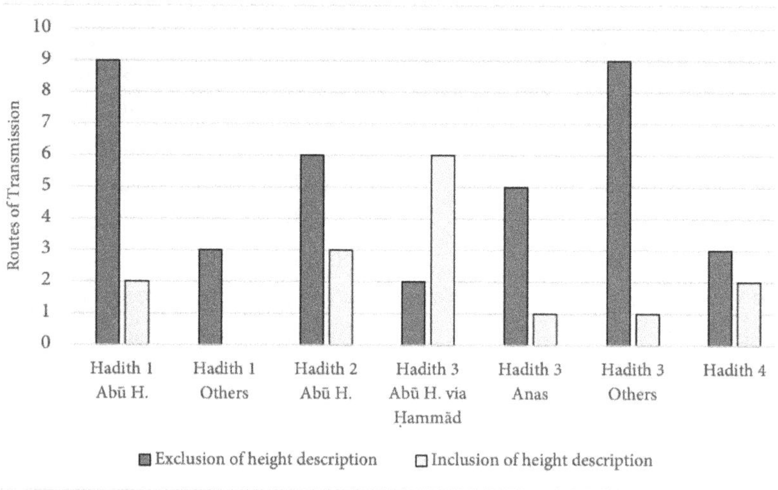

Figure 10: Summary of the hadith analysis vis-à-vis prophetic hadith. (The general findings are listed with more specificity in Appendix 2.)

To be sure, this analysis is not based on sheer numbers. Many of the routes that exclude the height description are transmitted through seminal students of the common source (e.g., al-Aʿraj → Abū Hurayra). Hence, the routes are also being "weighed."[109] Taking a bird's-eye view of prophetic hadith on the subject, it is clear that a preponderance of narrators, among whom many are "weighty" sources, do not transmit the two contentious clauses. This

108 Al-Samʿānī, *Qawāṭiʿ al-adilla*, 1:405; see also al-Suyūṭī, *Tadrīb al-rāwī*, 5:124. For a relevant discussion on preferring a hadith due to its inclusion in the two *Ṣaḥīḥ*s, see al-Nuʿmānī, *al-Imām Ibn Mājah*, 108–109; ʿAbd al-Rashīd al-Nuʿmānī, *Imām Ibn Mājah awr ʿilm-e ḥadīth* (Karachi: Aṣaḥḥ al-Maṭābiʿ, n.d.), 242–243.

109 Here I am borrowing Dr. Iftikhar Zaman's usage of the term. See Iftikhar Zaman, "The Science of *Rijāl* as a Method in the Study of Hadiths," *Journal of Islamic Studies* 5, no.1 (1994): 3.

preponderance is true for the stratum of the Companions (the height clause is arguably transmitted via Abū Hurayra alone), and it is predominantly the case for the second stratum of transmission, such as the narrators from Abū Hurayra. This fact decreases the epistemic value of the contentious words, thus making them far more probabilistic (*ẓannī*) in terms of their transmission. The point here, it should be emphasized, is not that the preponderance argument alone is sufficient to deem the contentious words unreliable. Rather, the epistemic value of the contentious words decreases substantially because of the preponderance argument, which then impacts how these words will fare when weighed against the empirical objections.

Another vital question is the status of textual omissions (*ikhtiṣār*) in hadith. When multiple routes of a single hadith are compared and some routes contain fewer clauses than others, is this a case of (1) convenient abridgment, (2) deliberate omission due to a defect (*ʿilla*), or (3) the fact that the common source (e.g., Abū Hurayra) relayed the hadith to some students without the addition? Moreover, was the hadith compiler (e.g., al-Bukhārī) responsible for the omission, or was it a narrator somewhere along the chain?

It was common for hadith compilers to truncate hadith in their collections (e.g., Hadith 2 via al-Firyābī). Muslim and Abū Dāwūd, for instance, explained that they intentionally truncated hadith in their respective collections.[110] The impetus to truncate a hadith partly stems from a desire to cite only parts of a hadith that are relevant to a given chapter or to remove material that they considered objectionable.[111]

---

110 Muslim, "Introduction," in *al-Musnad al-ṣaḥīḥ*, 1:3; Abū Dāwūd, *Risālat al-Imām Abī Dāwūd al-Sijistānī ilā Ahl Makka*, in Abū Ghudda, *Thalāth rasāʾil fī ʿilm muṣṭalaḥ al-ḥadīth* (Beirut: Dār al-Bashāʾir al-Islāmiyya, 2005), 32.

111 Anwar Shāh Kashmīrī argues that al-Bukhārī would truncate parts of a hadith in his *Ṣaḥīḥ* that he considered objectionable. See Kashmīrī, *Fayḍ al-Bārī*, 1:370; cf. Muḥammad al-Khanbarjī, "Ikhtiṣār al-matn wa-manhaj al-Imām al-Bukhārī fīhi" (PhD diss., University of Jordan, 2010), 153–161. For over a dozen examples of hadith scholars like Mālik, al-Bukhārī, Muslim, and al-Nasāʾī truncating parts of a hadith due to inherent defects, see Muḥammad ʿAwwāma, *Ḥadhf ṭaraf min al-ḥadīth al-wāḥid ikhtiṣāran aw iʿlālan* (Jeddah: Dār al-Minhāj, 2017), 10–32; see also Saʿīd Bāshanfar, *Manhaj al-Imām al-Bukhārī fī ʿarḍ al-ḥadīth al-maʿlūl fī al-Jāmiʿ al-ṣaḥīḥ* (Jeddah: Dār Ibn Ḥazm, 1436 AH), 44ff.

Muslim narrates that the Prophet (ṣ) was asked about fasting on a Tuesday, to which he responded, "I was born on that day." Muslim explains that the phrase "and Thursdays" was originally included, but he deliberately omitted it because he believed it was an error in transmission.[112] However, when multiple hadith collections record the textually omitted version from the same narrator (e.g., Hadith 3.2 via al-Awzāʿī), we can rule out the possibility of abridgement by the hadith compiler. Although hadith compilers did truncate hadith, textual omissions were mostly caused by a narrator during the process of transmission.[113] If it is proven that narrators themselves did not omit the words but rather differed on what their teacher transmitted,[114] a comparison between both versions would be conducted to determine the most accurate wording based on a host of factors.[115] For the most part, the hadith analyzed earlier appear to fall under this category.

---

112 *Ṣaḥīḥ Muslim*, no. 1162 (197).

113 See Ibn Ḥajar, *Fatḥ al-Bārī*, 13:248; Mastūra al-Muṭayrī, "al-Tanāẓur wa-l-tabāyun bayna ikhtiṣār al-ḥadīth wa-taqṭīʿihi ʿinda al-muḥaddithīn," *Majallat al-Dirāsa al-Islāmiyya* 13, no. 89 (2018): 430. This phenomenon is closely tied to the practice of paraphrased transmission. See Sulaymān al-Saʿūd, "Ikhtiṣār al-ḥadīth wa-atharuhu fī al-ruwāh wa-l-marwiyyāt: dirāsa waṣfiyya taḥlīliyya," *Majallat al-Jāmiʿa al-Islāmiyya* 9, no. 183 (1439 AH): 185–192. Al-Muṭayrī's claim that hadith compilers did not omit words from a hadith is undermined by numerous examples that suggest the contrary. See, for instance, Ibn Ḥajar, *Fatḥ al-Bārī*, 1:382; Ibn Rajab, *Fatḥ al-Bārī*, 2:105; Saʿīd Buwāʿina, "Ikhtiṣār al-riwāya ʿinda al-muḥaddithīn: dirāsa taʾṣīliyya," *Majallat Jāmiʿat al-Shāriqa* 17, no. 1 (2020): 668–670.

114 See Mujīr al-Khaṭīb, *Maʿrifat madār al-isnād*, 2:340–375, esp. 350 and 370.

115 Shaykh Yūnus Jawnpūrī offers a valuable observation concerning a trustworthy narrator's isolated transmission of a hadith (*tafarrud*). He states that a transmitter either narrates a hadith alone without any opposition (*mukhālafa*), like the hadith on actions being based on intentions, or he conflicts with others in his transmission. In the second situation, his isolated addition can be one of clear conflict, such as Jābir's hadith on the funeral prayer of the martyrs of Uḥud, or *he simply narrates what others do not* (*ʿadam al-dhikr*), such as Abū Hurayra's hadith on the repetition of words in the call to prayer (*tarjīʿ*). In the second situation, experts base their judgment regarding the acceptance of the addition on considerations that may tip the scale either in the narrator's favor or against him; this scenario is famously known as the addition of a reliable transmitter (*ziyādat al-thiqa*) [personal communication with Shaykh Jawnpūrī's

Prioritizing the contentions

In Jamīl Farīd's opinion, as stated in his recent study, the conflict between science and scripture has thus far reached an impasse: both sides are equally probabilistic, and therefore no judgment can be made until further corroboration tips the scale in favor of one side.[116] However, his analysis missed a crucial part of the debate: the positive evidence (i.e., our current knowledge of physics and biology, which suggests that humans could not have been anywhere near ninety feet tall). His study focuses solely on the fossil records (the negative evidence), which he contends only suggest the absence of evidence to support the hadith. He completely overlooks the positive evidence, which engenders no less, if not more, epistemic value compared to the archaeological evidence. Furthermore, as mentioned earlier, fossil records suggest that human height has not been decreasing in a linear fashion. Hence, the archaeological evidence does not merely prove the absence of evidence, as Jamīl Farīd presumes, but rather it constitutes evidence that conflicts with the notion of a gradual decrease in height.

Building upon Jamīl Farīd's analysis, by adding the positive evidence into the equation and revisiting the role of the archaeological evidence, a case can be made to prioritize these objections while accepting versions of the hadith that do not contain the two

---

student on January 4, 2020]. On the status of textual additions (*ziyādāt*), see al-Suyūṭī, *Tadrīb al-rāwī*, 3:319–331; Jonathan A.C. Brown, "Criticism of the Proto-Hadith Canon: al-Dāraquṭnī's Adjustment of the *Ṣaḥīḥayn*," *Journal of Islamic Studies* 15, no. 1 (2004): 7–11, 23–24. On the invariable acceptance of textual omissions over textual additions, see Pavel Pavlovitch, *The Formation of the Islamic Understanding of Kalâla in the Second Century AH (718–816 CE): Between Scripture and Science* (Leiden: Brill, 2016), 38.

116 Farīd, *Athar al-ʿilm al-tajrībī*, 141. Muḥyī al-Dīn al-Samarqandī writes that for the purposes of approximation, when hadith and empirical evidence are placed on the epistemic scale, hadith from the two *Ṣaḥīḥ*s and scientific laws engender the same epistemic value. Therefore, all things being equal, at a point of conflict, one is expected to suspend judgment on the matter. See al-Samarqandī, *Naqd matn al-ḥadīth*, 241. In our case, there are multiple factors that affect the epistemic value of both sides, so it is not a straightforward one-to-one comparison.

contentious clauses regarding Adam's height as sixty cubits and the gradual decrease of human height. It should also be recalled that al-Muṭahhar al-Maqdisī (d. ca. 390 AH) writes, "The sound opinion is that he was [tall] like a *saḥūq* (enormous) date palm. How many a date palm is shorter than human height; it [a date palm] is considered a *saḥūq* when it exceeds that [i.e., average human height]. The words 'sixty cubits' are possibly explanatory comments of a transmitter—and God knows best."[117] Likewise, Ibn Fūrak negated any hadith that described Adam as taller than ordinary human height.[118] Therefore, the proposal that both clauses should be dismissed is not entirely unprecedented. After I presented this research to Jamīl Farīd, he agreed that the scientific and archaeological contentions are weightier than he had initially estimated in his study, and he believes that prioritizing these contentions in the manner described here is a hermeneutically sound approach.[119]

As we encountered earlier, Yūnus Jawnpūrī prioritized the archaeological concerns vis-à-vis the gradual decrease of height by suggesting that the second clause specifically was a narrator insertion into the text of the hadith. Jawnpūrī argues that Abū Hurayra, the sole Companion from whom this phrase is reliably narrated, learned of the idea of people's height gradually decreasing from Kaʿb al-Aḥbār, who was well-versed in Biblical and Jewish traditions.[120] Subsequent transmitters from Abū Hurayra then mistakenly attributed it to the Prophet (ṣ).[121] As noted in our analysis, I have come across only two

117 Al-Maqdisī, *Kitāb al-Bad' wa-l-tārīkh*, 3:22.

118 Ibn Fūrak, *Mushkil al-ḥadīth*, 54–55.

119 Personal communication, June 25, 2021.

120 Adrienne Mayor explains how the idea that humans were originally giants gradually decreasing in height gained traction among the ancient Greeks. See Mayor, *First Fossil Hunters*, 199–202; Romano and Avanzini, "Skeletons of Cyclops and Lestrigons," 135. Some have attempted to extrapolate the idea of a gradual decline in human height based on calculations from the Bible and the Talmud. See Bondeson, *Cabinet of Medical Curiosities*, 75, 85.

121 Shaykh Yūnus Jawnpūrī mentions, "This is clearly a mistake. Abū Hurayra learned this from the *isrāʾīliyyāt*, for he used to sit with and learn from Kaʿb al-Aḥbār" (cited from his unpublished notes on *Ṣaḥīḥ al-Bukhārī*—partially published as *Nibrās al-sārī*—and via personal communication with two of his

routes of transmission for the second clause: (1) Maʿmar → Hammām → Abū Hurayra → the Prophet (ṣ) and (2) Kulthūm b. Muḥammad → ʿAṭāʾ al-Khurāsānī → Abū Hurayra → the Prophet (ṣ).[122] The second chain is unreliable because Kulthūm b. Muḥammad has been impugned and ʿAṭāʾ al-Khurāsānī did not hear from Abū Hurayra.[123] That leaves only the route of Hammām from Abū Hurayra. Hammām is unanimously accepted as a reliable transmitter, but ʿAlī b. al-Madīnī notes that in several instances he conflicted with other reliable transmitters from Abū Hurayra. Imam Aḥmad mentions that he was isolated in some of his reports.[124] Due to Hammām's occasional conflict with other transmitters and isolated transmission, the contemporary hadith scholar Ṣalāḥ al-Dīn al-Idlibī opines that one is encouraged to scrutinize hadith that Hammām solely transmitted.[125]

Abū Isḥāq Kaʿb b. al-Mātiʿ, better known as Kaʿb al-Aḥbār, was a Yemeni Jew who lived during the Prophet's lifetime but never met him and only accepted Islam after his death.[126] Kaʿb al-Aḥbār was a point of reference for a number of Companions for Jewish scriptural narratives, but these Companions had no qualms about openly

---

students). Elsewhere he said, "Abū Hurayra erred in this hadith." See Akram al-Nadwī, annotation on *al-Farāʾid fī ʿawālī al-asānīd wa-ghawālī al-fawāʾid* (Beirut: Dār al-Bashāʾir al-Islāmiyya, 2015), 80. Therefore, according to Jawnpūrī, "people have since been decreasing" are Abū Hurayra's words, which he learned from Kaʿb al-Aḥbār. By extension, given that later transmitters often confused the words of Abū Hurayra and those of Kaʿb, it is possible that the inclusion of this phrase in the prophetic hadith occurred later. Jawnpūrī even entertained the possibility that the given words were from Hammām rather than Abū Hurayra, since the former had exposure to earlier scripture, particularly through his brother Wahb b. Munabbih [personal communication with Jawnpūrī's student, January 4, 2020].

122 *Ṣaḥīḥ al-Bukhārī*, no. 3326 and *Musnad Isḥāq*, no. 433, respectively.

123 See the editor's comments in al-Ṭabarānī, *Musnad al-Shāmiyyīn*, no. 2332, 2369; cf. Abū Ḥudhayfa, *Anīs al-sārī*, 2:1606.

124 ʿUthmān b. Abī Shayba, *Suʾālāt ʿUthmān b. Muḥammad b. Abī Shayba li-l-Imām ʿAlī b. al-Madīnī* (Cairo: al-Fārūq al-Ḥadītha, n.d.), 39, no. 74; al-Mizzī, *Tahdhīb al-kamāl*, 30:300.

125 Al-Idlibī, *Manhaj naqd al-matn*, 388.

126 Ibn Ḥajar al-ʿAsqalānī, *al-Iṣāba fī tamyīz al-ṣaḥāba* (Beirut: Dār al-Kutub al-ʿIlmiyya, 1994) 5:481–484. Abū Shahba, *al-Isrāʾīliyyāt wa-l-mawḍūʿāt*, 100–105.

disagreeing with him.[127] There were instances where later narrators incorrectly attributed this material to the Prophet (ṣ),[128] particularly in the case of Abū Hurayra via Kaʿb. The Successor Busr b. Saʿīd (d. 100 AH) said that he attended the lectures of Abū Hurayra, who would narrate from both the Prophet (ṣ) and Kaʿb al-Aḥbār. When Abū Hurayra left, Busr overheard some of the attendees confusing the hadith of the Prophet (ṣ) with the statements of Kaʿb and vice versa.[129] Hadith scholars were aware of this phenomenon.[130] By no means should this be taken as an attack on Abū Hurayra's integrity as a transmitter.[131] A detailed analysis of Abū Hurayra's hadith corpus demonstrates that most of his reports are corroborated by other

127 See, for instance, the statement of Ḥudhayfa b. al-Yamān in Ibn Ḥajar, *al-Iṣāba,* 5:484.

128 ʿUmar Falāta, *al-Waḍʿ fī al-ḥadīth* (Damascus: Maktabat al-Ghazālī, 1981), 1:330–332.

129 Muslim b. al-Ḥajjāj, *Kitāb al-Tamyīz* (Riyadh: Sharikat al-Ṭibāʿa al-ʿArabiyya, 1982), 175, no. 10. There were students in those gatherings, like Busr b. Saʿīd in this case, who took note of this phenomenon. After citing this episode, Ibn Rajab al-Ḥanbalī (d. 795 AH) writes, "The discussion would grow too long if we were to mention the prophetic hadith that have been deemed defective because they were in fact suspended [i.e., non-prophetic] reports (*mawqūf*) either from ʿAbd Allāh b. Salām or Kaʿb [al-Aḥbār] that some of the transmitters confused with prophetic hadith." See Ibn Rajab, *Fatḥ al-Bārī*, 3:410; cf. Muḥammad ʿAwwāma, *Hal fī ḥadīth « khalaqa Allāh al-turba yawm al-sabt » ishkāl?* (Jeddah: Dār al-Minhāj, 2016), 29–31; al-Khaṭīb, *Maʿrifat madār al-isnād*, 1:322.

130 For instance, Abū Hurayra narrates from the Prophet (ṣ) that throughout the day Gog and Magog dig through the barrier behind which they are trapped, leaving a small space for the following day only to find it then restored to its original state. This will continue until they say, "God willing (*in shāʾ Allāh*)," after which they will find the wall as they left it and break through. Ibn Kathīr comments that probably Abū Hurayra heard this from Kaʿb, from whom something similar is related, and then subsequent narrators incorrectly ascribed it to the Prophet (ṣ). See Ibn Kathīr, *Tafsīr al-Qurʾān al-ʿAẓīm*, 5:177. For other examples, see al-Idlibī, *Manhaj naqd al-matn*, 94; Farīd, *Athar al-ʿilm al-tajrībī*, 268; al-Bukhārī, *al-Tārīkh al-kabīr*, 1:413; see also Ibn Khuzayma's comments on the hadith that Adam was created on a Friday in *Ṣaḥīḥ Ibn Khuzayma*, no. 1729.

131 For a study on how Ibrāhīm al-Nakhaʿī, ʿĪsā b. Abān, and other Ḥanafīs viewed Abū Hurayra's hadith, see al-Turkumānī, *Dirāsāt*, 229ff.; ʿAbd al-Munʿim al-ʿIzzī, *Difāʿ ʿan Abī Hurayra* (Beirut: Dār al-Qalam, 1981), 237ff.; Bīrī Zādah, *al-Fatḥ*

Companions.[132] Furthermore, a study of Abū Hurayra's life and works undercuts any reason to cast aspersions on his integrity.[133]

Inclusion in the *Ṣaḥīḥayn*

It is true that al-Bukhārī and Muslim narrate versions of this hadith in their *Ṣaḥīḥ*s that include the description of sixty cubits. However, Jamīl Farīd counters that at times the inclusion of a hadith in their respective collections does not necessitate that they invariably deem every word of that report to be authentic.[134] There are instances where al-Bukhārī[135]

---

al-Raḥmānī sharḥ Muwaṭṭa' al-Imām Muḥammad b. al-Ḥasan al-Shaybānī, fol. 166r, Istanbul: Yūsuf Āghā Library, manuscript no. 338.

132 According to Ḍiyā' al-Raḥmān al-A'ẓamī's study, there are fewer than three hundred authentic and fair hadith transmitted via Abū Hurayra that are not corroborated by other Companions. See Ḍiyā' al-Raḥmān al-A'ẓamī, "Abū Hurayra fī ḍaw' marwiyyātihi bi-shawāhidihā" (MA thesis, Jāmi'at al-Mālik 'Abd al-'Azīz, 1973), section *ḥā'*. Usman Ghani also examined the hadith in Abū Hurayra's corpus and compared them with the reports of other prolific Companions (*mukthirūn*). He concludes that Abū Hurayra's hadith are vastly corroborated. See Usman Ghani, "'Abū Hurayrah' a Narrator of Ḥadīth Revisited," (PhD diss., University of Exeter, 2011), 215–266. The common number of hadith attributed to Abū Hurayra, based on Ibn Ḥazm's estimation, is 5,374. However, Aḥmad Shākir explains that this number is problematic; the actual number is much smaller. See Aḥmad Shākir, *al-Bā'ith al-ḥathīth* (Riyadh: Maktabat al-Ma'ārif, 1996), 2:507–512.

133 There are numerous books written on Abū Hurayra as a transmitter of hadith, such as 'Ajāj al-Khaṭīb's *Abū Hurayra rāwiyat al-Islām* and 'Abd al-Mun'im al-'Izzī's *Difā' 'an Abī Hurayra.*

134 Farīd, *Athar al-'ilm al-tajrībī*, 137.

135 For instance, al-Bukhārī narrates a hadith via Ibn Numayr [no. 6251] where the Prophet (ṣ) instructs a person to repeat his prayer. This version of the report concludes with the words, "then rise until you are composed while sitting." Al-Bukhārī follows this report with a version from Abū Usāma that states, "[then rise] until you are standing upright." Ibn Ḥajar explains that al-Bukhārī cited Abū Usāma's version to highlight that Ibn Numayr's version is erroneous. See Ibn Ḥajar, *Fatḥ al-Bārī*, 2:279, 11:37. For an example related to the chain of transmission, see Ibn Ḥajar, 9:401. At times, al-Bukhārī intentionally cites a hadith in an unrelated chapter to allude to a defect in the chain or text of a hadith. See 'Imād al-Ṣimādī, "Maqāṣid al-Bukhārī fī riwāyat al-aḥādīth fī ghayr maẓānnihā," (PhD diss., Jāmi'at al-'Ulūm al-Islāmiyya al-'Ālamiyya [Amman],

and Muslim,[136] explicitly or implicitly, highlight parts of a hadith or its chain in their *Ṣaḥīḥ*s that they themselves deem problematic.[137] This is not to suggest that all the hadith on the present subject should be dismissed in their entirety. Rather, the contentious clauses are part of lengthier hadith, and therefore, the conversation revolves around parts of a hadith the inclusion of which narrators have differed over. As we have seen in our analysis, al-Bukhārī and Muslim relate both versions of the hadith, both with and without the contentious clauses. Under a hadith in *Ṣaḥīḥ Muslim* on the topic of breastfeeding, Muḥammad Taqī ʿUthmānī writes:

> In view of these three considerations [i.e., comments from Ibn al-Ṣalāḥ, Ibn Ḥajar, and Shabbīr ʿUthmānī], if experts like al-Ṭaḥāwī and [Abū Bakr] Ibn al-ʿArabī critique this part of the hadith, they are not dissenting from the consensus of the Muslim community [on the authenticity of the two *Ṣaḥīḥ*s], particularly when they do not object to the reliability of the crux of the hadith. They accept the hadith overall, but they object to the addition

2017), 75–97. For a detailed study on al-Bukhārī's method of presenting defective chains and texts in his *Ṣaḥīḥ*, see Saʿīd Bāshanfar, *Manhaj al-Imām al-Bukhārī*.

136 For instance, Muslim (no. 119) narrates from Anas that when the verse "O you have believed, do not raise your voices above the voice of the Prophet (ṣ) . . ." (Q. 49:2) was revealed, Thābit b. Qays stayed at home, remorseful for having raised his voice. The Prophet (ṣ) was unaware of Thābit's whereabouts, so he asked Saʿd b. Muʿādh about him. Muslim then cites three routes from Anas that retell the same incident without any mention of Saʿd b. Muʿādh. Muslim himself makes it a point after citing each route to reiterate this omission. Through this practice, Muslim points out that the mention of Saʿd in this hadith is defective. The reason for this is that the verse was revealed in 9 AH, whereas Saʿd b. Muʿādh had passed away years earlier, in 5 AH. See Ibn Kathīr, *Tafsīr*, 7:367; ʿAwwāma, "Introduction," in *Muṣannaf Ibn Abī Shayba* (Jeddah: Dār al-Qibla, 2006), 1:107–108. Muslim's method of critiquing hadith is well known among experts in the field. Ḥamza al-Malībārī provides an early treatment of this methodology in his rejoinder to Rabīʿ al-Madkhalī entitled "*Abqariyyat al-Imām Muslim fī tartīb aḥādīth Musnadihi al-ṣaḥīḥ*" and in a subsequent publication "*Mā hākadhā tūridu yā Saʿd al-ibil*."

137 To be clear, these examples are meant to support the validity of the maxim cited above, not to suggest that al-Bukhārī and Muslim are applying this method of critique to this specific hadith.

mentioned solely by ʿAbd Allāh b. Abī Bakr to the exclusion of other transmitters. No one disagrees on the occurrence of errors by some narrators in some hadith in the two *Ṣaḥīḥ*s.[138]

Early hadith scholars operated within a two-tiered system. Depending on the topic of the hadith, they applied different critical lenses: hadith on theology and law required more stringency than hadith about the end times and exhortations. This dichotomy might help to explain why a hadith included in an authentic compilation could nevertheless be subjected to scrutiny. This does not mean that hadith masters like al-Bukhārī and Muslim dropped their standards in certain chapters; rather, this two-tiered system was a "systematic feature of Sunni Hadith transmission and criticism."[139] Defending the presence of a questionable hadith in the *Ṣaḥīḥ*, Ibn Ḥajar explains that al-Bukhārī was less stringent in this case because the hadith is of an exhortatory nature (*targhīb*).[140]

As some contemporary scholars note, the gates of reexamining the authenticity of certain words in hadith featured in the two *Ṣaḥīḥ*s are open for researchers who take into consideration the necessary measures (e.g., working within the framework of the disciplines of hadith). This was the approach of latter-day hadith experts like Ibn al-Qaṭṭān (d. 628 AH), al-Nawawī, Ibn Taymiyya, and Ibn Ḥajar.[141] There is a sizable list of classical scholars whose fidelity to the two *Ṣaḥīḥ*s did not prevent them from critically engaging with certain words they believed required further scrutiny. Obviously, this critical engagement only occurred when the requisite criteria for such scrutiny were met.[142]

---

138 ʿUthmānī, *Takmilat Fatḥ al-Mulhim*, 4:44–45. See also Ibn Taymiyya, *Majmūʿ fatāwā*, 18:352.

139 Brown, *Misquoting Muhammad*, 256–260; Brown, *Hadith*, 270–271.

140 Ibn Ḥajar al-ʿAsqalānī, *Hudā al-sārī*, 1:440; see also Abū Ghudda, "Annotations," in al-Laknawī, *Ẓafar al-amānī bi-sharḥ Mukhtaṣar al-Sayyid al-Sharīf al-Jurjānī* (Aleppo: Maktab al-Maṭbūʿāt al-Islāmiyya, 1416 AH), 185.

141 Zaryūḥ, *al-Muʿāraḍāt al-fikriyya*, 2:730.

142 Muḥyī al-Dīn al-Samarqandī provides a list of scholars—based on the work of the Ibāḍī writer Saʿīd al-Qanūbī—who engaged critically with the two *Ṣaḥīḥ*s, starting with the contemporaries of al-Bukhārī and Muslim until modern-day scholars. See al-Samarqandī, *Naqd matn al-ḥadīth*, 345; cf. Jonathan A.C.

Within the framework of conflict resolution, prioritization (*tarjīḥ*) is the process of preferring the epistemically weightier of two conflicting sides. At least four hadith reference the height of Adam as sixty cubits, but a preponderance of their narrators disagree on the inclusion of the contentious clauses, thus lowering their epistemic status. Jamīl Farīd's initial analysis led him to argue that both sides are equal in strength, obviating the possibility of rendering a definitive judgment. His assessment, however, overlooked crucial scriptural and empirical details. By filling in these gaps, there is scope to prioritize the empirical contentions concerning both clauses—a conclusion which Farīd later conceded is sound. Yūnus Jawnpūrī argued that the second clause, on the gradual decrease in human height, appears to be a narrator insertion due to archaeological concerns and its isolated transmission. One could, therefore, interpret the height of Adam figuratively without needing to address the resulting problems with the second clause. It is important to note that these arguments are not without precedent in Islamic intellectual history.

Nonetheless, one may feel uncomfortable with or unconvinced by the conclusions of prioritization. At the same time, one may be hard pressed to offer a suitable method to resolve the contentions. This brings us to the third step in the process of conflict resolution: the suspension of judgment.

Brown, *The Canonization of al-Bukhārī and Muslim* (Leiden, Brill, 2007), 300ff.; Shabbīr Aḥmad 'Uthmānī, *Mabādi' 'ilm al-ḥadīth wa-uṣūluhu* (Beirut: Dār al-Bashā'ir al-Islāmiyya, 2011), 160; al-Turkumānī, *Dirāsāt*, 127–134; 'Ubayd Allāh Sindhī, *Risāla fī muṣṭalaḥ al-ḥadīth* (Karachi: Ghulām Muṣṭafā, n.d.), 19.

SUSPENSION OF JUDGMENT (*TAWAQQUF*)

After exhausting all the hermeneutic techniques in their toolbox, scholars would simply suspend judgment in the face of uncertainty or conflicting evidence.[1] Scholars have taken noncommittal stances in nearly every Islamic discipline,[2] such as the theological debate on the createdness of the Qur'ān,[3] the meaning of the imperative verb (*amr*) in legal theory,[4] and the ritual purity of a mule's saliva in law.[5] The field of hadith is no exception. Abū ʿĪsā al-Tirmidhī (d. 279 AH) asked his teachers al-Bukhārī and al-Dārimī (d. 255 AH) to choose

1 Abū Ḥāmid al-Ghazālī, *al-Mustaṣfā* (Beirut: Dār al-Kutub al-ʿIlmiyya, 1993), 1:36–37. See also ʿUmar al-Faramāwī, "Min masālik al-muḥaddithīn wa-l-uṣūliyyīn fī al-taʿāmul maʿa mukhtalif al-ḥadīth," *Majallat Dār al-Iftāʾ al-Miṣriyya* 19, no. 1 (2014): 57–60.

2 Those who suspend judgment on rulings in the presence of conflicting evidence are given the interesting epithet *wāqifa* (lit. those who stand still or stop). See Najm al-Dīn al-Ṭūfī, *Sharḥ Mukhtaṣar al-Rawḍa* (Beirut: Muʾassasat al-Risāla, 1987), 1:391. Historically, seekers of sacred knowledge were taught to say "I do not know" (*lā adrī*) as part of their educational upbringing. See Muḥammad ʿAwwāma, *Maʿālim irshādiyya li-ṣināʿat ṭālib al-ʿilm* (Jeddah: Dār al-Minhāj, 2011), 333–336.

3 On those who took a noncommittal stance on the issue of the createdness of the Qur'ān and its repercussions in the field of hadith, see Abū Ghudda, *Masʾalat khalq al-Qurʾān* (Aleppo: Maktab al-Maṭbūʿāt al-Islāmiyya, n.d.), 10–15; see also Badr al-Ghāmidī, *al-Tawaqquf fī al-ʿaqīda: dirāsa fī al-manhaj wa-l-masāʾil wa-l-asbāb ʿinda ahl al-sunna* (London: Takwīn li-l-Dirāsāt wa-l-Abḥāth, 2016).

4 ʿAbd al-Raḥmān ʿAzzāz, "al-Tawaqquf fī al-masāʾil al-uṣūliyya fī dalālāt al-alfāẓ" (MA thesis, Jāmiʿat al-Imām Muḥammad b. Saʿūd al-Islāmiyya, 2005), 104–118.

5 For a list of legal rulings regarding which Abū Ḥanīfa took a noncommittal stance, see Ibn ʿĀbidīn, *Radd al-muḥtār ʿalā al-Durr al-mukhtār* (Damascus: Dār al-Thaqāfa al-Islāmiyya, 2001), 3:515–519.

the most reliable route of a particular hadith, but they did not pass any judgment.[6] More specifically, scholars often suspended judgment on hadith whose meanings they were unable to understand or the contentions surrounding which they were unable to resolve. When asked about the hadith on washing a utensil that had been licked by a dog, Mālik said, "This hadith has come [to us], but I do not know what the reality of it is."[7]

In a similar vein, Ibn Ḥajar al-ʿAsqalānī took a noncommittal stance on the last part of the hadith on Prophet Adam's height. He comments on the words "humankind has since been decreasing," saying:

> This is problematized by the remnants of previous civilizations, like the cities of Thamūd. Their homes indicate that their stature was not excessively tall as would be demanded by the succession [of gradual decrease mentioned in the second clause]. They were undoubtedly a nation of old, and the timespan between them and Adam is less than that between them and the first of this nation. *I have yet to come upon a solution to this problem.*[8]

The Syrian hadith expert Muḥammad ʿAwwāma reiterates Ibn Ḥajar's archaeological concerns and adds to this the artifacts and skeletal remains of the pharaohs. Despite the size of the pyramids, the pharaohs were the same size as humans today, which does not correspond to the successive decrease mentioned in the hadith.[9]

---

6 Abū ʿĪsā al-Tirmidhī, *al-ʿIlal al-kabīr* (Beirut: ʿĀlam al-Kutub, 1988), no. 11; see also Jalāl Rāghūn, *al-Tawaqquf ʿinda al-muḥaddithīn: dirāsa ta'ṣīliyya taṭbīqiyya*, 2014).

7 Saḥnūn, *al-Mudawwana al-kubrā* (Beirut: Dār al-Kutub al-ʿIlmiyya, 1994), 1:115; Umar F. Abd-Allah, *Mālik and Medina: Islamic Legal Reasoning in the Formative Period* (Leiden: Brill, 2013), 117. Adopting a noncommittal approach to problematic inquiries and hadith was commonplace for Imām Aḥmad. See, for instance, Isḥāq al-Marwazī, *Masā'il al-Imām Aḥmad b. Ḥanbal wa-Isḥāq b. Rāhawayh* (Medina: al-Jāmiʿa al-Islāmiyya, 2002), nos. 520, 732, 826, 861; Abū Bakr al-Khallāl, *Ahl al-milal wa-l-ridda wa-l-zanādiqa* (Riyadh: Maktabat al-Maʿārif, 1996), 1:73; Ibn Ḥajar al-ʿAsqalānī, *Badhl al-māʿūn fī faḍl al-ṭāʿūn* (Riyadh: Dār al-ʿĀṣima, n.d.), 148.

8 Ibn Ḥajar, *Fatḥ al-Bārī*, 6:366–367.

9 Muḥammad ʿAwwāma, *Min ṣiḥāḥ al-aḥādīth al-qudsiyya* (Jeddah: Dār al-Minhāj, 2011), 111–112. I would like to thank my three-year-old daughter for

As a rejoinder to Ibn Ḥajar's comments, the Indian hadith scholar and commentator Zakariyyā Kāndhlawī (d. 1982) provides an analogy to resolve the archaeological difficulties involved in accepting a literal interpretation of the hadith. He writes that the creation can be likened to the growth of a child. The era between Adam and Noah was its childhood, the era from Noah to Abraham was its adolescence, and the era after that was the beginning of its adulthood. A person's height increases more quickly from infancy to the end of childhood than in any other stage of life. In a similar but inverse manner, the major decrease in the height of humankind occurred in the era between Adam and Noah, with the subsequent decrease occurring much more slowly. Kāndhlawī argues, therefore, that it is possible that by the time of Thamūd, human height had already decreased substantially, which explains the size of their dwellings.[10] Two centuries earlier, al-Amīr al-Ṣanʿānī (d. 1768) offered a response similar to that of Kāndhlawī.[11] Likewise, Ismāʿīl al-ʿAjlūnī (d. 1749) responded to Ibn Ḥajar's concerns by saying that the decrease in height was gradual for most of human history, which does not preclude the possibility that there were periods of "punctuated equilibrium."[12]

These arguments are problematized by scientific findings that height never decreased continuously in a linear fashion, not to mention the problems with assuming that such drastic physiological changes occurred in humans at a rapid pace.[13] The phenomenon of phyletic or insular dwarfism should not be presented to support the

---

unwittingly directing my attention to this reference.

10 Zakariyyā Kāndhlawī, *al-Abwāb wa-l-tarājim* (Beirut: Dār al-Bashā'ir al-Islāmiyya, 2012), 4:396.

11 Al-Amīr al-Ṣanʿānī, *al-Tanwīr sharḥ al-Jāmiʿ al-ṣaghīr* (Riyadh: Maktabat Dār al-Salām, 2011), 5:498.

12 Ismāʿīl al-ʿAjlūnī, *al-Fayḍ al-jāri bi-sharḥ Ṣaḥīḥ al-Bukhārī*, vol. 5, fol. 415r, MS Aḥmad Thālith.

13 Interestingly, al-Suyūṭī was asked, "Has it been transmitted that Adam and his immediate children were sixty cubits, the second generation forty, the third twenty, and the fourth seven?" He responds by saying that there is no scriptural proof for these specific numbers; what is transmitted is that Adam was sixty cubits and that his progeny has since been shrinking. See Jalāl al-Dīn al-Suyūṭī, *al-Ḥāwī li-l-fatāwī* (Beirut: Dār al-Fikr, 2004), 1:430.

type of rapid and drastic change that these scholars have posited. There were specific environmental conditions that resulted in rapid dwarfing among a subset of a certain species, such as a limitation of resources due to isolation on an island. The above claim, on the other hand, is that humans as a species collectively shrank dramatically in rapid succession. More importantly, mammals like dwarf elephants that shrank because of the "island rule" maintained the overall anatomical structure of their mainland ancestors.[14] A similar reduction in size among humans would require drastic physiological and anatomical changes for reasons explained in chapter two. At any rate, whether Ibn Ḥajar's concerns can be resolved need not detain us.[15] The focus here is to appreciate his willingness to suspend judgment in the face of an apparent conflict between scriptural data and empirical evidence.

Muhammad Zubayr Siddiqi (d. 1976) cites this hadith as an example of weak or forged hadith that made their way into the canonical hadith collections.[16] He bases this judgment on Ibn Ḥajar's reser-

---

14 On insular dwarfism, see Roberto Rozzi and Mark Lomolino, "Rapid Dwarfing of an Insular Mammal – The Feral Cattle of Amsterdam Island," *Science Report* 7 (2017); Alexandra van der Geer et al., "From Jumbo to Dumbo: Cranial Shape Changes in Elephants and Hippos During Phyletic Dwarfing," *Evolutionary Biology* 45 (2018); Ana Benítez-López et al., "The Island Rule Explains Consistent Patterns of Body Size Evolution in Terrestrial Vertebrates," *Nature Ecology & Evolution* (2021).

15 Shaykh Rāghib al-Ṭabbākh (d. 1951) mentions that he wrote an unpublished twenty-page treatise dealing with Ibn Ḥajar's comments. See al-Ṭabbākh, *Dhū al-Qarnayn wa-sadd al-Ṣīn* (Kuwait: Ghirās, 2003), 196. See also Zaryūḥ, *al-Muʿāraḍāt al-fikriyya*, 3:1432ff.

16 Dr. Muḥammad Zubayr Siddiqi was the author of seminal hadith works in English and other languages, such as *Hadith Literature: Its Origin, Development, Special Features & Criticism.* He was born ca. 1890 in Bihar, India. He received his initial Islamic education from his father Muḥammad Isḥāq, after which he studied at a local madrasa, and then he graduated from an *ʿāliya* madrasa in Rampur. After completing his M.A. at Patna University, he traveled to Cambridge University to complete his PhD under the supervision of Edward G. Browne (d. 1926). Shortly after his return to India, Siddiqi served as Sir Ashutosh Professor of Islamic Culture at Calcutta University where he headed the department from 1929 until his retirement in 1962. In addition to his expertise in hadith, Siddiqi was formally trained in traditional medicine

vations concerning the implications of this hadith.[17] However, Ibn Ḥajar's comments demonstrate that he merely took a noncommittal stance to understanding the hadith; he did not deem the hadith weak, let alone forged.

The contention that led Ibn Ḥajar to suspend judgment regarding the hadith was clearly not definitive (*qaṭʿī*). Yet, it was a strong enough reason for him not to dismiss the validity of the contention. Ibn Ḥajar only suspended judgment on the last part of the hadith (i.e., that the height of humankind has been decreasing since Adam). Given the concerns posed by a literal explanation of the first part of the hadith (i.e., the height of Adam being sixty cubits), there is scope to extend Ibn Ḥajar's approach and suspend judgment on that as well. As noted earlier, even al-Ḥasan al-Baṣrī stated that "God knows best which cubit [is intended]" in reference to Adam's height being sixty cubits.[18]

(*ḥikma*); he published and edited works in the field. He passed away in 1976. On the life and works of Muḥammad Zubayr Siddiqi, see Masʿūd Ḥasan, "Wafayāt: Dāktar Muḥammad Zubayr Ṣiddīqī," *Maʿārif* 177, no. 4 (1976).

17 Muhammad Zubayr Siddiqi, *Hadith Literature: Its Origin, Development & Special Features* (Cambridge: Islamic Texts Society, 1993), 114–115.

18 Yaḥyā b. Sallām, *Tafsīr*, 2:815.

MIRACLES AND THE LAWS OF NATURE

Epistemic requirements for miracles

There are many hadith that defy our current understanding of science, such as the splitting of the moon during the Prophet's time, and yet these are accepted literally without offering a figurative reading.[1]

1 On the splitting of the moon, see al-Muṭīʿī, *Tawfīq al-Raḥmān*, 746–761. Shāh Walī Allāh writes concerning the splitting of the moon, "It is not necessary that the moon was physically split. It was possibly something similar to smoke, shooting stars, and solar and lunar eclipses that appear in the horizon to the sight of people. To express these phenomena in Arabic, words are used that denote the occurrences themselves. The Qurʾān was revealed in the language of the Arabs. Akin to this is the report of ʿAbd Allāh b. Masʿūd, who requires no introduction: 'They were afflicted with a drought. Whenever they looked, they saw smoke in the sky. Concerning this incident, the verse "[Then watch thou] for the Day that the sky will bring forth a kind of smoke plainly visible" was revealed.'" See Dihlawī, *Taʾwīl al-aḥādīth*, 102–103. Ghulām Muṣṭafā Qāsimī, the editor of *Taʾwīl al-aḥādīth*, notes that in one manuscript of the work, this passage is prefaced with the words "someone who has familiarity with archaeology and the natural sciences said," which means that Dihlawī is quoting someone else's view. Qāsimī then adds that al-Kawtharī's criticism of Dihlawī is, therefore, misplaced (he is referring here to al-Kawtharī's statement that this was one of Dihlawī's anomalous views). See al-Kawtharī, *Ḥusn al-taqāḍī*, 248. In addition, Kashmīrī responds to the criticisms of a similar view found in Dihlawī's *al-Tafhīmāt al-ilāhiyya*. See Dihlawī, *al-Tafhīmāt al-ilāhiyya* (Bijnor: Madīna Barqī Press, 1936), 2:57; Anwar Shāh Kashmīrī, *al-ʿArf al-shadhī* (Beirut: Dār al-Turāth al-ʿArabī, 2004), 4:334. For an overview of the exchanges that ensued among Indian scholars in the nineteenth century as a result of Dihlawī's position, see Nūr al-Ḥasan, "al-Tafhīmāt al-ilāhiyya awr uskey tardīd wa-taʾyīd" *Maʿārif* 184, no. 4 (2009): 280–282.

Why then, one might ask, should the hadith about Adam's height be treated any differently? For one, to establish that an event interrupts the habitual course of nature (*kharq al-ʿāda*) requires epistemically satisfactory evidence. On the authentic but solitary hadith (*khabar wāḥid*) that God suspended the movement of the sun upon the request of Prophet Joshua (Yūshaʿ),[2] Bakhīt al-Muṭīʿī writes that technically God can allow the manifestation of miracles that break the course of observable natural laws; these miracles, however, follow subtle laws that humans fail to perceive.[3] Scholars have still opted to interpret this authentic hadith about Prophet Joshua in a way that conforms to observable realities, e.g., that it refers to an increase of blessings in time, not that the sun was physically suspended.[4] It is, therefore, better to interpret the hadith figuratively than to read it literally and assume that natural laws were altered to accommodate it.[5] Al-Muṭīʿī then adds, citing Fakhr al-Dīn al-Rāzī (d. 606 AH), that

2 *Ṣaḥīḥ al-Bukhārī*, no. 3214; *Ṣaḥīḥ Muslim*, no. 1747; Ibn Ḥajar, *Fatḥ al-Bārī*, 1:293.

3 In a similar vein, Shāh Walī Allāh writes, "Indeed, what is termed 'a break [in the course of nature]' (*kharq*) is part of the natural occurrences, but since its causes are rare and it occurs infrequently such that people do not expect its manifestation, it is termed 'a break.'" See Dihlawī, *Ta'wīl al-aḥādīth*, 101. For an analysis of al-Ghazālī's Ashʿarī stance on the reversibility of the natural order and Ibn Rushd's subsequent rejoinder in his *Incoherence of the Incoherence* vis-à-vis prophetic miracles, see Isra Yazicioglu, "Redefining the Miraculous: al-Ghazālī, Ibn Rushd and Said Nursi on Qur'anic Miracle Stories," *Journal of Qur'anic Studies* 13, no. 2 (2011). See also Ibn Ḥazm's critique of the Ashʿarī stance in *al-Fiṣal fī al-milal* (Cairo: Maktabat al-Salām al-ʿĀlamiyya, n.d.), 5:11–12; cf. Abū Bakr al-Bāqillānī, *Kitāb al-Bayān ʿan al-farq bayna al-muʿjizāt wa-l-karāmāt*, ed. Richard J. McCarthy (Beirut: al-Maktaba al-Sharqiyya/ Librairie Orientale, 1958), 50ff.

4 He analogizes this to the hadith in which the Prophet (ṣ) tells Abū Dharr that during sunset, the sun prostrates underneath the Divine Throne and then seeks permission to rise again (*Ṣaḥīḥ al-Bukhārī*, no. 3199). Given the scientific problems with accepting this hadith at face value, scholars have offered several figurative interpretations of it.

5 Shaykh Bakhīt al-Muṭīʿī's observations should not be written off as an attempt at pandering to a modernist audience. He was a staunch Ḥanafī Māturīdī judge and was vocal about his disagreement with reformist scholars like Muḥammad ʿAbduh and Rashīd Riḍā. The deputy to the last Ottoman Shaykh al-Islām,

the hadith on the sun retracting is ultimately a solitary report, and an occurrence of this magnitude requires recurrent mass transmission (*tawātur*) to be accepted.[6]

There is no credible archaeological evidence to prove that our ancestors were extraordinarily giant humans who then gradually decreased in height. Additionally, our current knowledge of physics and biology suggests that humans could not have been anywhere near ninety feet tall while maintaining their current physiological and anatomical properties. To establish anything that goes against these premises would require evidence that is epistemically acceptable in its transmission (*thubūt*) and semantic import (*dalāla*). In view of the analysis provided earlier, the hadith on the height of Adam—the contentious words specifically—and the gradual decrease of his progeny is particularly probabilistic (*ẓannī*) in both its transmission and semantic import. As such, resolving the inherent tension via the three-tiered method is hermeneutically more appropriate than accepting the hadith at face value. Contrast this with the splitting of the moon, which is definitive (*qaṭʿī*) in both its transmission and semantic import.[7] To opt for a figurative reading in this case is unwarranted. Comparing the two hadith is therefore a false equivalence—as is the case with similar scriptural texts.

Muḥammad Zāhid al-Kawtharī (d. 1952), praised al-Muṭīʿī and held him in high regard. See Khayr al-Dīn al-Ziriklī, *al-Aʿlām: qāmūs tarājim li-ashhar al-rijāl wa-l-nisāʾ min al-ʿArab wa-l-mustaʿribīn wa-l-mustashriqīn* (Beirut: Dār al-ʿIlm li-l-Malāyīn, 2002), 6:50; al-Kawtharī, *Maqālāt*, 139. On the life and thought of Bakhīt al-Muṭīʿī, see Quadri, *Transformations of Tradition*.

6 Al-Muṭīʿī, *Tawfīq al-Raḥmān*, 554, 569, cf. 474–478; al-Rāzī's comments were in the context of Prophet Solomon. See al-Rāzī, *Mafātīḥ al-ghayb*, 26:291. For al-Muṭīʿī's perspective on saintly wonders (*karāmāt*), see al-Muṭīʿī, 745; see also Ibn Baṭṭāl, *Sharḥ Ṣaḥīḥ al-Bukhārī*, 5:208–209. For an overview of shifting attitudes towards *karāmāt*, see Jonathan A.C. Brown, "Faithful Dissenters: Sunni Skepticism about the Miracles of Saints," *Journal of Sufi Studies* 1, no. 2 (2012).

7 On the transmission of the splitting of the moon via *tawātur*, see Muḥammad b. Jaʿfar al-Kattānī, *Naẓm al-mutanāthir min al-ḥadīth al-mutawātir* (Cairo: Dār al-Kutub al-Salafiyya, n.d.), 211–212. On the semantic import of this hadith, see al-Muṭīʿī, *Tawfīq al-Raḥmān*, 758. For a concise analysis of modern discussions on the splitting of the moon, see Farīd, *Athar al-ʿilm al-tajrībī*, 84–88.

During the twentieth century, reformist scholars like Rashīd Riḍā opined that only epistemically certain evidence can establish the interruption of nature's habitual course, thus limiting the number of acceptable miracles.[8] The last Ottoman Shaykh al-Islām Muṣṭafā Ṣabrī (d. 1954) engaged in a heated polemic with Egypt's modernist movement on the prophetic miracles and dedicated a significant portion of his magnum opus to critiquing their views.[9] Even traditional scholars were not uncritical in their acceptance of reports of this nature.[10] Advocating for a reasonable standard of verification, ʿAbd al-Fattāḥ Abū Ghudda (d. 1997) refutes al-Sakhāwī's effort to establish reports that the Prophet (ṣ) was greeted by a gazelle and writes, "These are flimsy, unreliable hadith that cannot be relied upon to establish an interruption in nature's habitual course. The implication of these reports can only be accepted via a preponderantly authenticated hadith (*al-ṣaḥīḥ al-rajīḥ*)."[11]

Earlier, I used the words "epistemically satisfactory" when describing the requirements for evidence proving a miraculous event so as not to limit them with the restrictive criteria of Riḍā while nonetheless maintaining the need for added caution advocated by traditional scholars. Moreover, reports on miracles or breaks in the habitual course of nature cannot be subsumed under one category; each incident deserves a relevant degree of scrutiny. For instance, the splitting of the moon was a major event that had universal implications, and therefore scholars were more exacting in their examination of it. It

---

8 Rashīd Riḍā, "Mas'alat inshiqāq al-qamar," *Majallat al-manār* 30 (1348 AH): 361; Maḥmūd Shaltūt, *Fatāwā* (Cairo: Dār al-Shurūq, 2001), 48–53; cf. Fahd al-Rūmī, *Manhaj al-madrasa al-ʿaqliyya*, 579ff.

9 See, for instance, Muṣṭafā Ṣabrī, *Mawqif al-ʿaql wa-l-ʿilm wa-l-ʿālam min Rabb al-ʿālamīn wa-ʿibādihi al-mursalīn* (Beirut: Dār Iḥyāʾ al-Turāth al-ʿArabī, 1981), 1:108–112.

10 Although the *sīra* genre generally applied more relaxed standards, later authors of the *sīra* and *shamāʾil* literature like al-Suyūṭī (d. 911 AH) went a step further and included miraculous events with unreliable chains that would not have otherwise been accepted, such as the revival of the Prophet's parents. See Brown, "Faithful Dissenters," 155.

11 Abū Ghudda, "Annotations," in Mullā ʿAlī al-Qārī, *al-Maṣnūʿ fī maʿrifat al-ḥadīth al-mawḍūʿ* (Beirut: Dār al-Bashāʾir al-Islāmiyya, 1994), 80.

is important to note that even a solitary hadith that is augmented by auxiliary factors enjoys increased epistemic value, while negative factors weaken it.[12] Interestingly, the Muʿtazilī theologian Qāḍī ʿAbd al-Jabbār (d. 415 AH) in his day allowed the use of solitary reports on prophetic miracles to attain inspiration (*li-yashraḥa ṣudūr al-muʾminīn*) but not to prove prophethood (*iḥtijāj*). These incidents, however, could have served this purpose during the Prophet's time without being massively transmitted thereafter.[13]

Implications of the hadith

Attempts to resolve the problems with the hadith on Adam's height by explaining it as a prophetic miracle (*muʿjiza*) fail to consider the broader implications of the hadith, not to mention the function of a prophetic miracle.[14] The hadith also states that subsequent generations were of a similar but gradually decreasing height, at which point the issue ceases to involve Adam alone. Furthermore, Ibn Hubayra had anticipated another implication when he raised the question about Adam's horse: if people were that large, everything else, such as their food, shelter, and personal effects, would have been proportionately sized.[15] One would, therefore, have to accept

---

12 See Ibn Ḥajar's discussion on strengthening the epistemic value of the Prophet's (ṣ) miracles that were transmitted through solitary lines of transmission in *Fatḥ al-Bārī*, 6:582.

13 ʿAbd al-Jabbār, *al-Mughnī fī abwāb al-tawḥīd wa-l-ʿadl* (Cairo: al-Dār al-Miṣriyya li-l-Taʾlīf wa-l-Tarjama, n.d.), 16:407. I would like to thank Mawlana Zeeshan Chaudri for directing me to this reference.

14 On the varying definitions and qualifications of a *muʿjiza*, see Muḥammad al-ʿUmarī, *al-Nubuwwa bayna al-mutakallimīn wa-l-falāsifa wa-l-ṣūfiyya* (Amman: Dār al-Fatḥ, 2015), 228–254. According to the theologians, among the criteria for something to be considered a *muʿjiza* is that it correspond to a prophet's proclamation (*daʿwā*), prove his truthfulness, and not be subject to challenge (*muʿāraḍa*). See al-ʿUmarī, 230–231.

15 Interestingly, latter-day scholars like Ibn al-Ḍiyāʾ (d. 854 AH) attribute a statement to Kaʿb al-Aḥbār that a grain of wheat in the time of Adam was the size of an ostrich egg, and then it continued to shrink over time until it became the size it is today. See Ibn al-Ḍiyāʾ, *Tārīkh Makka al-musharrafa* (Beirut: Dār al-Kutub al-ʿIlmiyya, 2004), 45; cf. Paul Hershon, *A Talmudic Miscellany*

that the entire world was drastically different, a very serious claim that lacks the requisite evidence. Interestingly, Tāj al-Dīn al-Subkī (d. 771 AH) mentions that the experts opine that saintly wonders and even prophetic miracles cannot occur so regularly that they are confused for the habitual course of nature.[16]

A detailed study on resorting to "*kharq al-ʿāda*" (a break in the habitual course of nature) to resolve conflicts between hadith and empirical realities is imperative but beyond the scope of the present study. The decision to opt for this route was based on a wide array of considerations, such as the epistemic value of the hadith in question, the weight of the objection, and the implications of such an interpretation. Scholars were not monolithic in the application of this principle. Consider two distinct approaches in grading the hadith that says, "The Prophet (ṣ) came out of his house one day and had in his hand two books with the names of the people of Paradise and the people of Hell—their names and the names of their fathers and their tribes."[17] The physical impossibility of the Prophet (ṣ) holding two massive books that would have weighed "several *qirāṭs*" led Shams al-Dīn al-Dhahabī (d. 748 AH) to conclude that the report is highly objectionable (*munkar jiddan*). On the other hand, Ibn Ḥajar responds that the weight of the books is irrelevant because it was a tremendous miracle (*muʿjiza ʿaẓīma*).[18] Books that iden-

(Oxford: Routledge, 2000), 30. We saw earlier al-Ḥakīm al-Tirmidhī's view that "a grain of wheat was the size of an ox's kidney, one pomegranate enough for ten people, and a bunch of grapes equal in size." See al-Tirmidhī, *Nawādir al-uṣūl*, 1:140.

16 Al-Subkī, *Ṭabaqāt al-Shāfiʿiyya al-kubrā*, 2:319.

17 *Jāmiʿ al-Tirmidhī*, no. 2141.

18 Al-Dhahabī, *Mīzān al-iʿtidāl*, no. 5325; Ibn Ḥajar, *Lisān al-Mīzān*, no. 4990. Mūsā Lāshīn's (d. 2009) reservations regarding the classical interpretation of the Prophet's (ṣ) words "I can see you from behind me when you bow and prostrate (in prayer)" as a miracle is also instructive for our purposes. See Mūsā Lāshīn, *Fatḥ al-Munʿim sharḥ Ṣaḥīḥ Muslim* (Cairo: Dār al-Shurūq, 2002), 2:584; cf. Muḥammad al-Ithyūbī, *Sharḥ Sunan al-Nasāʾī* (Riyadh: Dār al-Miʿrāj, 1996), 10:249. Abū al-Walīd al-Bājī's (d. 474 AH) controversial interpretation of the writing of the Ḥudaybiyya treaty as a prophetic miracle and the immediate debacle that ensued is an insightful example of this form of interpretation, on the one hand, and the scholarly reservations about its implications, on the

tify problematic hadith solely based on their content state that, in principle, anything that is falsified by empirical realities (*shawāhid ṣaḥīḥa*)—like the mythic ʿŪj, son of ʿŪq (King Og), who allegedly stood a towering 3,333 cubits tall!—is a forgery, without entertaining the possibility of an interruption in the habitual course of nature.[19]

Anticipating a paradigm shift

The current state of scientific and archaeological research could hypothetically change in light of yet unknown developments in physics, anatomical studies, and archaeology—a paradigm shift, as Thomas Kuhn would put it.[20] However, the mere possibility of a shift in research does not warrant a dismissal of the present concerns, as they are grounded in sufficient evidence to be taken seriously. Ibn Ḥajar acknowledged an archaeological problem with the hadith predicated on a contention that was not epistemically conclusive. Preponderant conviction alongside the possibility of error is satisfactory evidence that goes beyond mere speculation. The Ḥanafī legal theorist Ṣadr al-Sharīʿa al-Maḥbūbī (d. 747 AH) writes that epistemic certainty can also refer to that which precludes substantive errors (*ʿilm al-ṭumaʾnīna*).[21] Ḥāfiẓ al-Dīn al-Nasafī (d. 710 AH) relates a definition of *ṭumaʾnīna* as "conviction in the preponderance of accuracy with the possibility of doubt or error."[22] Abū Ḥāmid al-Ghazālī expounds on the epistemic value of empirical propositions known through

---

other. See Abū al-Walīd al-Bājī, *Taḥqīq al-madhhab* (Riyadh: ʿĀlam al-Kutub, 1983), chap. 4; Ibn Ḥajar, *Fatḥ al-Bārī*, 503–504. For a detailed account of what Joel Blecher dubs "the Bājī affair," see Joel Blecher, *Said the Prophet of God: Hadith Commentary across a Millennium* (Oakland: University of California Press, 2018), 21–29.

19 Ibn al-Qayyim, *al-Manār al-munīf*, 76.

20 For an accessible and relevant introduction to the philosophy of science, the ideas of Thomas Kuhn (d. 1996), and the debate between scientific realism and antirealism, see James Ladyman, *Understanding Philosophy of Science* (London: Routledge, 2002), 11–91, 93–123, and 129–262, respectively.

21 ʿUbayd Allāh al-Maḥbūbī, *al-Tawḍīḥ sharḥ al-Tanqīḥ* (Cairo: Maktabat Ṣabīḥ, n.d.), 1:248.

22 Ḥāfiẓ al-Dīn al-Nasafī, *Kashf al-asrār* (Beirut: Dār al-Kutub al-ʿIlmiyya, n.d.), 2:6–7; cf. ʿAlāʾ al-Dīn al-Bukhārī, *Kashf al-asrār* (Beirut: Dār al-Kitāb al-Islāmī,

observing reoccurring events in the world (*tajrībiyyāt*; e.g., the law of gravity). He writes that these propositions provide certainty to those who experience them, with varying degrees of certainty depending on one's knowledge and experience.[23]

Limits of scientific inquiry

It goes without saying that simply because certain hadith are studied through a scientific lens to better gauge their veracity, it does not follow that scripture in its entirety can be put under a microscope. A holistic understanding of this issue requires a critical exposé of scientism and its various strands—ontological, epistemic, and rationalistic—which, again, goes beyond the current study.[24] Studying every hadith through a strictly naturalistic lens overlooks the fact that providing guidance on scientific matters was not the raison d'être of the Prophet's (ṣ) mission. Shāh Walī Allāh Dihlawī (d. 1762) makes the apt observation that the prophets did not occupy themselves with matters unrelated to "the refinement of the soul and governing of the community unless it was incidental."[25]

It should be abundantly clear by now that Muslim scholars have explored diverse empirical methods to determine the precise grading of hadith, but they acknowledged the parameters of such exploration. Hadith address issues pertaining to three domains: the seen (*shahāda*), the unseen (*ghayb*), and the relatively unseen (e.g., the causes of celestial phenomena).[26] Scripture and science may overlap

n.d.), 2:364; Muḥammad Hindū, al-*Kulliyyāt al-tashrīʿiyya* (Amman: al-Maʿhad al-ʿĀlamī li-l-Fikr al-Islāmī, 2016), 110–111.

23 Al-Ghazālī, *al-Mustaṣfā*, 1:36–37. For an insightful discussion on the epistemic frameworks related to the ethico-legal status of seeking out medical treatment, see Omar Qureshi and Aasim Padela, "When Must a Patient Seek Healthcare? Bridging the Perspectives of Islamic Jurists and Clinicians into Dialogue," *Zygon* 51, no. 3 (2016): 605–617.

24 See, for instance, Malik, *Atheism and Islam*, 23–24.

25 Yūsuf Bannūrī, *Maʿārif al-sunan sharḥ Jāmiʿ al-Tirmidhī* (Karachi: H.M. Saʿīd, 1992), 5:349; Dihlawī, *Ḥujjat Allāh al-bāligha* (Beirut: Dār al-Jīl, 2005), 1:158.

26 For an insightful elaboration of Ibn Taymiyya's demarcation of the seen and unseen realms—both are inherently perceptible although the human senses are

in the first and, to some degree, the third categories.[27] Scriptural matters of the unseen—such as the angels and the afterlife—are by definition supra-rational and metaphysical, and ipso facto, they should not and cannot be weighed on the limited scales of science and human reason,[28] as they occupy different domains.[29] Venturing into this territory with a naturalistic worldview is destined for failure. Ibn Khaldūn compares the incommensurability of such a worldview with the realm of the seen by providing the following analogy, "One might compare it with a man who sees a scale in which gold is being weighed and wants to weigh mountains in it."[30]

---

veiled from perceiving the latter—and his exposé on the scope of inferring the unseen on the basis of the seen (*qiyās al-ghā'ib 'alā al-shāhid*), see El-Tobgui, *Ibn Taymiyya on Reason and Revelation*, 230–238, 277–285.

27 Al-Samarqandī, *Naqd matn al-ḥadīth*, 9, 208; al-Muṭī'ī, *Tawfīq al-Raḥmān*, 474–478. An example of the third category is the hadith "[Thunder is] the angel of Allah who is responsible for the clouds. In his hand is a whip of fire with which he drives the clouds to where Allah commands." See *Jāmi' al-Tirmidhī*, no. 3117; for a study of this hadith, see Farīd, *Athar al-'ilm al-tajrībī*, 144–150, 186–194.

28 Noson Yanofsky provides fascinating and engaging examples of the limits of reason and its related areas: science, technology, logic, and mathematics. See, for instance, the chessboard and dominos example, the ship of Theseus, Russell's paradox, and the problem of induction; Yanofsky, *The Outer Limits of Reason: What Science, Mathematics, and Logic Cannot Tell Us* (Cambridge, MA: The MIT Press, 2013), 2, 31, 85, 236, and 340–345, respectively.

29 On the role of science in Islamic discourse and in hadith particularly, see al-Samarqandī, *Naqd matn al-ḥadīth*, 189–215; Farīd, *Athar al-'ilm al-tajrībī*, 29ff.

30 The full quote is as follows: "The intellect, indeed, is a correct scale. Its indications are completely certain and in no way wrong. However, the intellect should not be used to weigh such matters as the oneness of God, the other world, the truth of prophecy, the real character of the divine attributes, or anything else that lies beyond the level of the intellect; that would mean to desire the impossible. One might compare it with a man who sees a scale in which gold is being weighed, and wants to weigh mountains in it." See Ibn Khaldūn, *Muqaddimah,* 3:38; Ibrahim Kalin, *Reason and Rationality in the Qur'ān* (Amman: Kalam Research & Media, 2015), 6–7.

CONCLUSION

Debates on the conflict between scripture and science, with particular focus on hadith, have increased at an unprecedented rate in the modern age. Premodern scholars were by no means unaware of this conflict. They dealt with similar debates and, in the process, provided a coherent methodology for analogous situations. This methodology involves a critical appraisal of both sides of the conflict along a spectrum of epistemic value, followed by an evaluation based on a three-tiered model—harmonization (*jamʿ*), prioritization (*tarjīḥ*), and suspension of judgment (*tawaqquf*)—to determine the best course of action. The hadith describing Prophet Adam's height as sixty cubits and humankind's subsequent decrease in height has been placed in the spotlight due to scientific and archaeological concerns that a literal reading of it poses. In this study, we examined how scholars from different periods in Islamic history have proposed to deal with this hadith through the three-tiered model of conflict resolution.

First, contemporary scholars like al-Muʿallimī and Kashmīrī maintain that the hadith could easily be describing a metaphysical phenomenon: Adam's height was sixty cubits only in Paradise, not during his stay on earth. This is the most balanced interpretation on the subject. Reconciling this interpretation with the last part of the hadith (i.e., a gradual decrease in human height) has been a bone of contention, with some proposing an alternative reading of the relevant words and others suggesting that a narrator insertion is at play.

Second, based on a holistic analysis of all the chains and versions of the hadith, we learn that the majority of transmitters do not narrate the passages concerning the height of Adam and his progeny. With a decrease in the epistemic value of these passages due to the conflicting routes of transmission, a case can be made to give credence to the scientific and archaeological concerns while accepting versions of the hadith that do not include the description of height and subsequent gradual decrease. Alternatively, as Jawnpūrī contends, we can opt for a hybrid approach: namely, prioritize the empirical objections by dismissing only the last part of the hadith while maintaining that Adam's height was sixty cubits in Paradise. These proposals may come across as novel. However, this interdisciplinary perspective is not without merit, though it requires further exploration.

Third, Ibn Ḥajar—an exceptionally qualified scholar—had no qualms about suspending judgment regarding a gradual decrease in height due to his inability to answer an archaeological conundrum in accepting it at face value. Understandably, many have taken his cue. Although Ibn Ḥajar suspended judgment only on the last part of the hadith (concerning a gradual decrease in human height), given the underlying motivation for his noncommittal stance, we can reasonably extend his reservation to the issue of Adam's height—or other ostensibly problematic hadith for that matter.

While researching this topic, I could not help but notice the extensive commentary on determining the antecedent of the pronoun in the words "God created Adam in *His/his* image" found in some versions of the hadith, with little debate on the present topic (although a minority of voices did participate in such a debate). Ḥamūd al-Tuwayjirī (d. 1992) wrote an entire monograph on the question of whether Adam was created in God's form.[1] Centuries earlier, the Mālikī jurist Aḥmad al-Fayyūmī (d. 1101 AH) wrote a treatise on the stages of Adam's creation up until his demise.[2] Al-Fayyūmī dedicates only a few lines to the archaeological contention on the hadith, in which he briefly quotes Ibn Ḥajar's comments mentioned above. The situation today is shifting, whereby the limelight has fallen on the question of Adam's height without serious controversy on the pronoun debate (His vs. his). This observation highlights

1 See Ḥamūd b. ʿAbd Allāh al-Tuwayjirī, *ʿAqīdat ahl al-īmān fī khalq Ādam ʿalā ṣūrat al-Raḥmān* (Riyadh: Dār al-Liwāʾ, 1989).

2 See Aḥmad al-Fayyūmī, *al-Qawl al-tāmm fī bayān aṭwār sayyidinā Ādam ʿalayhi al-ṣalāh wa-afḍal al-salām* (Riyadh: King Abdul Aziz Library, manuscript no. 684).

the influence that shifting paradigms and new discoveries have on scholarly debates and commentary.[3]

Students of Islamic intellectual history may notice that premodern discussions on reason and revelation centered around issues that may not immediately resonate with modern concerns (e.g., the divine attributes). Sharif El-Tobgui, however, observes in reference to Ibn Taymiyya that

> the underlying problematic remains, in significant ways, very much the same. Whether it is the issue not precisely of reason and revelation but, say, of science and revelation or, for instance, the tension between sacralized and secularized visions of law and government, which has been a particularly troubling issue for Muslims in the modern period, the root of all these issues can be traced to the deeper lying tensions with which Ibn Taymiyya grappled when confronting the delicate question of the relationship between reason and revelation in his own day.[4]

An examination of scholarly efforts to resolve the tension surrounding the hadith on Adam's height yields results that go beyond this specific issue. It serves as an excellent case study of the hermeneutic techniques that both classical and modern scholars have applied when they found themselves at the crossroads of science and scripture.

And Allah knows best.

---

3 On how hadith commentaries have responded to contemporary issues, see Blecher, *Said the Prophet of God*, 13–18; Muhammad Qasim Zaman, *The Ulama in Contemporary Islam: Custodians of Change* (Princeton: Princeton University Press, 2002), 40–50.

4 El-Tobgui, *Ibn Taymiyya on Reason and Revelation*, 19–20.

AFTERWORD

The Problem of the *Isrāʾīliyyāt*

Jonathan A.C. Brown

A storyteller in medieval Baghdad would enthrall his audiences with tales of the prophets, filled with detail. In one session, he recounted that "the name of the wolf that ate Joseph was such and such." An audience member objected, "Joseph was not eaten by a wolf," to which the storyteller replied, "Then that is the name of the wolf that *didn't* eat Joseph."[1]

Details are important. They can make rules and principles more effective and stories more interesting. But the quest for details is a fraught one. Though they can bring generalities to life in particular, details can also lead to a betrayal of the themes or principles they are meant to color. Already in the 900s CE, Muslim scholars and litterateurs writing about the history of the world sullied their pages with stories about how Black Africans had been cursed with blackness and servitude because their ancestor, Ham, had not honored his father Noah properly. These Muslim scholars cited authorities like Wahb b. Munabbih (d. 114/732) and Ibn Isḥāq (d. 150/767). Later, master scholars like Ibn al-Jawzī (d. 597/1201) and Jalāl al-Dīn al-Suyūṭī (d. 911/1505) corrected this mistake, showing that these reports were not sound or reliable.[2] Wahb b. Munabbih wrote some of the earliest

1 Ibn al-Jawzī, *Kitāb al-Quṣṣāṣ wa-l-mudhakkirīn*, ed. Merlin Swartz (Beirut: Dar El-Machreq, 1986), 112.

2 Jalāl al-Dīn al-Suyūṭī, *Rafʿ shaʾn al-ḥubshān*, ed. Muḥammad ʿAbd al-Wahhāb Faḍl (Cairo: self-published, 1991), 270–271; Jonathan A.C. Brown, *Slavery and Islam* (London: Oneworld, 2019), 121–123.

surviving texts in the Islamic tradition. But they often lacked *isnad*s entirely and simply copied Jewish and Christian lore. Despite being an important hadith scholar, Ibn Isḥāq was criticized for narrating material indiscriminately from the People of the Book.[3]

Reacting to the hadith (or, more properly, the section found in several hadiths) in which the Prophet (ṣ) states that "God created Adam in his image," the Egyptian Islamic modernist Maḥmūd Abū Rayya (d. 1970) blamed *isrāʾīliyyāt*—stories from the Bible or the Jewish and Christian traditions surrounding it.[4] In his opinion, the idea of saying that God had created man in His own image plainly contradicted the Qurʾān's declaration that "there is nothing like unto Him" (Q. 42:11) and had clearly seeped into the Islamic tradition from the Biblical book of Genesis (1:27) (to be clear, Muslim scholars understood this hadith in a variety of ways, from denying that humans could grasp its meaning to pointing to the grammatical rule that the "his" must refer to Adam, not God).[5] Abū Rayya was a severe critic of the Sunni hadith tradition, and his book *Lights on the Muhammadan Sunna* (*Aḍwāʾ ʿalā al-sunna al-muḥammadiyya*, 1958) prompted at least eight book-length rebuttals. What is interesting is that, as Muntasir Zaman points out in this excellent work, Shaykh Yūnus Jawnpūrī (d. 2017) argued that at least part of this hadith was a product of the *isrāʾīliyyāt* (namely, the section on humans decreasing in height, the subject of the book). And what is even more interesting is that this argument has as a precedent no less than one of the founders of the Ashʿarī school of theology, Ibn Fūrak (d. 406/1015), who posited that narrations suggesting that Adam exceeded ordinary human height were taken from "the narratives of the Old Testament" (*aḥādīth al-Tawrāh*).[6]

---

3 Ibn Ḥajar al-ʿAsqalānī, *Tahdhīb al-Tahdhīb*, ed. Muṣṭafā ʿAbd al-Qādir ʿAṭā, 12 vols. (Beirut: Dār al-Kutub al-ʿIlmiyya, 1994), 9:37.

4 G.H.A. Juynboll, *The Authenticity of the Tradition Literature: Discussions in Modern Egypt* (Leiden: Brill, 1969), 88, 133.

5 Saʿd al-Dīn al-Taftāzānī et al., *Majmūʿat al-ḥawāshī al-bahiyya ʿalā sharḥ al-ʿAqāʾid al-Nasafiyya*, 4 vols. (Cairo: Maṭbaʿat Kurdistān, 1329 AH), 1:97; Burhān al-Dīn Ibrāhīm al-Bayjūrī, *Ḥāshiyat al-Imām al-Bayjūrī ʿalā Jawharat al-tawḥīd*, ed. ʿAlī Jumʿa (Cairo: Dār al-Salām, 2006), 159.

6 Ibn Fūrak, *Mushkil al-ḥadīth*, 55.

Blaming what one might call the blight of the *isrāʾīliyyāt* has been common in the last century, particularly among Muslim scholars of a modernist or reformist bent. These reports drawn from the scripture and lore of the People of the Book, in particular the Rabbinic Jewish tradition (i.e., the Talmudic commentaries on Jewish law and *midrash*, a body of stories linked to the Biblical text), are blamed for insinuating anthropomorphic, superstitious, unscientific, and misogynistic ideas into the Islamic community. We find this accusation voiced by more influential and respected reformist scholars like Muḥammad ʿAbduh (d. 1905), Rashīd Riḍā (d. 1935), and Shaykh al-Azhar Maḥmūd Shaltūt (d. 1964), as well as by more controversial figures like Muḥammad Tawfīq Ṣidqī (d. 1920) and Abū Rayya (who even penned a small article entitled "Kaʿb al-Aḥbār, the First Zionist").[7] According to these critics, *isrāʾīliyyāt* had been ushered into Muslim books and minds primarily to flesh out the Qurʾān's narratives about Biblical prophets like Noah and Moses. Their stories are referred to elliptically in the Qurʾān but are not laid out in full detail, the assumption being that the Prophet's audience knew the stories already and needed only to be reminded of the lessons they imparted or to have specific points corrected. Shaltūt in particular railed against how, instead of protecting the Muslim masses from the influences of foreign religious traditions that had been adulterated beyond repair, Muslim scholars of *tafsīr* had allowed these reports to pollute their understanding of the Qurʾān's narratives and teachings.[8]

It is ironic that the suspicion that Islamic modernists and reformists felt towards *isrāʾīliyyāt* was inspired by the works of Ibn Taymiyya (d. 728/1328) and his student Ibn Kathīr (d. 774/1373), two scholars

7 Juynboll, *Authenticity of the Tradition Literature*, 28–30; Daniel Brown, *Rethinking Tradition in Modern Islamic Thought* (Cambridge: Cambridge University Press, 1996), 89; Rashīd Riḍā, *Majallat al-Manār* 26, no. 1 (1925): 73–79; *Majallat al-Manār* 27, no. 7 (1926): 539ff.; *Majallat al-Manār* 27, no. 8 (1926): 610–618; *Majallat al-Manār* 28, no. 1 (1927): 27; *Majallat al-Manār* 32, no. 10 (1931): 753; Muḥammad Abū Rayya, *Aḍwāʾ ʿalā al-Sunna al-Muḥammadiyya* (Cairo: Dār al-Taʾlīf, 1958), 169; Abū Rayya, "Kaʿb al-Aḥbār huwa al-ṣahyūnī al-awwal," *al-Risāla wa-l-Riwāya* 665 (1946).

8 Maḥmūd Shaltūt, *Fatāwā* (Cairo: Dār al-Shurūq, 1983), 5556.

usually associated with the more conservative Salafi movement.[9] But it is not surprising. Just as Ibn Taymiyya's books had first been published by the reformist Riḍā and the arch-Salafi Muḥammad Nāṣir al-Dīn al-Albānī (d. 1999) had begun his journey in hadith studies after reading Riḍā's reformist *Manār* journal, on the topic of *isrāʾīliyyāt* we find the not uncommon intersection of "Salafi" and Islamic modernist discourse. But Ibn Taymiyya had clearly been onto something. A Yemeni Sufi scholar who penned a famous history of the Bāʿalawī scholars and Sufis of Yemen, al-Shillī (d. 1093/1682), had mentioned that it should not be permitted to read books like stories of the prophets (*qaṣaṣ al-anbiyāʾ*) in the mosque, nor should one read books with mostly forged contents or those taken from sources that cannot be relied upon from among the People of the Book.[10]

Although, as we will see, the subject goes back to the Prophet's time, the origins of the term *isrāʾīliyyāt* seems to come from the tenth century CE. Tottoli first locates the term in al-Masʿūdī's (d. 345/956) *Murūj al-dhahab* to denote reports drawn from Jewish writings and with somewhat fantastical contents.[11] We then see it used several times in the influential pietistic treatise *Qūt al-qulūb* of the Basran Sufi Abū Ṭālib al-Makkī (d. 386/996).[12] The tension surrounding the issue of *isrāʾīliyyāt*, however, goes all the way back to the Prophet's (ṣ) time and the revelation of the Qur'ān. Where should Muslim scholars turn for the context and details that seemed necessary to mine the Qur'ān's stories of previous prophets and draw lessons from them?

The richest body of material lay in the Old Testament. But there a great challenge loomed: the Qur'ān had made clear that previous communities such as the Christians and the Jews had altered or seriously misunderstood previous revelations (Q. 2:75, 4:46, 5:13, 5:41). If the text had been altered, how could Muslims know that any material they drew from it was reliable? Do not seek knowledge of

---

9 Tottoli, "Origin and Use of the Term *Isrāʾīliyyāt*," 202.

10 Muḥammad b. Abī Bakr al-Shillī Bāʿalawī, *al-Mushriʿ al-rawī fī manāqib al-sāda al-kirām Āl Abī ʿAlawī*, 2 vols. (Cairo: al-Maṭbaʿa al-ʿĀmira, 1319 AH), 1:145.

11 Tottoli, "Origin and Use of the Term *Isrāʾīliyyāt*," 194–195.

12 Abū Ṭālib al-Makkī, *Qūt al-qulūb*, 2 vols. in 1 (Cairo: Maṭbaʿat al-Anwār al-Muḥammadiyya, 1985), 1:141, 2:224.

religion from the People of the Book, warned Ibn ʿAbbās (d. 67/686-7) in *Ṣaḥīḥ al-Bukhārī*, since God had told Muslims that previous communities had altered (or misrepresented, *baddalū*) their books and since, anyway, the Qur'ān had superseded them.[13]

This message is also found in numerous hadiths, though most are unreliable. In a sound hadith in *Ṣaḥīḥ al-Bukhārī*, the Prophet (ṣ) warns his followers not to believe or disbelieve reports they hear from the People of the Book but just to say, "We believe in God and His messengers." As the Prophet (ṣ) adds in another narration generally considered sound as well, this is so that any truth these reports might contain not be rejected and any falsehood they convey not be believed.[14] A set of similar but weak hadiths found in the *Musnad* of Imām Aḥmad (d. 241/855) and all narrated through the Kufan scholar al-Shaʿbī (d. ca. 105/723) describe ʿUmar showing the Prophet (ṣ) a book he had gotten from Jews in Medina. The Prophet (ṣ) replies:

> And will you all stumble into this, O son of al-Khaṭṭāb? By Him who holds my soul in His hand, I have come to you all with what is pure and untainted, so do not ask them about anything, that they might tell you something true and you disbelieve it, or tell you something false and you believe it. By God, if Moses were alive, he would have no choice but to follow me.[15]

In a hadith with an incomplete *isnād*, al-Dārimī (d. 255/868) reports a hadith in which the Prophet (ṣ) warns that it would suffice for Muslims to be led astray for them to turn away from their prophet and their book to the prophets and books of others.[16]

These warnings, however, are complicated by the best-known teaching from the Prophet (ṣ) on the contents of books like the Bible and the lore of the People of the Book. In a hadith considered *ṣaḥīḥ*, the Prophet (ṣ) said, "Narrate from the Children of Israel without

---

13 *Ṣaḥīḥ al-Bukhārī*: kitāb al-shahādāt, bāb lā tas'al ahl al-shirk.

14 Ibid.: kitāb al-tafsīr, bāb Sūrat al-Baqara qūlū āmannā . . . ; *Sunan Abī Dāwūd*: kitāb al-ʿilm, bāb riwāyat ḥadīth ahl al-kitāb; al-Baghawī, *Sharḥ al-sunna*, 1:269.

15 The first narration and the second al-Arnā'ūṭ deems weak due to the presence of Mujālid in the *isnāds*. The third al-Arnā'ūṭ deems weak due to the presence of Jābir al-Juʿfī in the *isnād*; see *Musnad Aḥmad*, 23:349; 22:468; 25:198.

16 *Sunan al-Dārimī*: introductory chapters, bāb man lam yara kitābat al-ḥadīth.

discomfort," a clause that was sometimes included along with the Prophet's (ṣ) well-known warning against misrepresenting him and sometimes with the additional clause, "for among them there are wonders (*aʿājīb*)."[17]

How do we understand these seemingly contrasting instructions? In some ways, the tension here is a consequence of a revelation that comes to seal the messages and stories brought by previous prophets, both affirming the core themes of their monotheistic call and also correcting the errancies and willful disobedience of the communities that formed around them. The Qurʾān came both to confirm "what was before it" (Q. 3:3, etc.) and to correct those who had misinterpreted it, namely, the People of the Book. On the one hand, the Prophet (ṣ) came in the tradition of those messengers who had come before, confirming what they had brought. But at the same time, the Qurʾān replaced some of the rites and rulings of previous communities (such as observing the Sabbath and some dietary rules), giving the "Umma of Muḥammad" its own ritual markings (*shaʿāʾir*): the day of Islam's communal prayer was Friday, not Saturday or Sunday; Mecca, not Jerusalem, was the final direction of prayer. The Prophet (ṣ) loved to conform his dress and conduct with the traditions of the People of the Book until the revelation instructed him otherwise.[18] Sometimes the Qurʾān made the break unmistakable. In the Book of Deuteronomy, if a couple divorces and then the woman marries another man, she can never again marry her first husband, for "this would be detestable in the eyes of the Lord" (Deut. 24:1–4). By contrast, the Qurʾān rules that a man who

---

17 *Sunan* of Abū Dāwūd: kitāb al-ʿilm, bāb al-ḥadīth ʿan Banī Isrāʾīl. This clause is found in *Ṣaḥīḥ al-Bukhārī*: kitāb aḥādīth al-anbiyāʾ, bāb mā dhukira ʿan Banī Isrāʾīl, and it is also graded as *ṣaḥīḥ* by al-Sakhāwī, al-Suyūṭī, Ibn Rajab, Shuʿayb al-Arnāʾūṭ, and al-Albānī; Shams al-Dīn al-Sakhāwī, *al-Maqāṣid al-ḥasana*, ed. Muḥammad ʿUthmān Khisht (Beirut: Dār al-Kitāb al-ʿArabī, 2004), 192; al-Suyūṭī, *al-Jāmiʿ al-ṣaghīr*, no. 3691; Ibn Rajab al-Ḥanbalī, *Ahwāl al-qubūr*, ed. ʿĀṭif Ṣābir Shāhīn (Mansoura: Dār al-Ghad al-Jadīd, 2005), 71; *Musnad Aḥmad*, 11:25, 16:125; al-Albānī, *Silsilat al-aḥādīth al-ṣaḥīḥa* (Riyadh: Maktabat al-Maʿārif, 1996), 6:1028–1030, no. 2926.

18 *Ṣaḥīḥ Muslim*: kitāb al-faḍāʾil, bāb sadl al-Nabī (ṣ).

had repudiated his wife cannot remarry her *unless* she has married another man (Q. 2:230).

So as Muslims sought to mine meaning and lessons from the Qur'ān's stories of previous prophets, the heritages of those earlier communities and their scriptures were sources at once temptingly valid and dangerously compromised. Muslim scholars resolved the tension in a variety of ways. Some scholars like Ibn Ḥajar (d. 852/1449) and al-Zarkashī (d. 792/1394) held that the permission to narrate from the People of the Book was granted later in the Prophet's career after the initial risk of confusing such material with the Prophet's teachings had receded (but, as Ibn 'Abbās's warning shows, this danger *never really* receded).[19] Imam al-Shāfi'ī (d. 204/820) explains that what the Prophet (ṣ) allowed was accepting reports from individuals among the People of the Book even though one could not be sure about those individuals' truthfulness (*ṣidq* or *kadhib*). But he did not allow narrating reports that were known to be false from the People of the Book.[20] This explanation comes from al-Shāfi'ī's own book, the *Risāla*, but in another explanation from a later text attributed to al-Shāfi'ī, he says that the Prophet (ṣ) meant that Muslims can narrate reports they have heard from Jewish tradition even though they are impossible or impermissible in the context of Islam, though this was not a mandate to propagate lies.[21]

The problem with these explanations is that they simply kick the can down the road. Yes, one can narrate from the People of the Book provided one is not narrating falsehoods; but the whole problem is that the unreliability of their scriptures and traditions means that one

---

19 Ibn Ḥajar al-'Asqalānī, *Fatḥ al-Bārī sharḥ Ṣaḥīḥ al-Bukhārī*, ed. 'Abd al-'Azīz bin Bāz and Ayman Fu'ād 'Abd al-Bāqī, 14 vols. (Beirut: Dār al-Kutub al-'Ilmiyya, 1997), 6:617; Badr al-Dīn al-Zarkashī, *al-La'ālī al-manthūra fī al-aḥādīth al-mashhūra*, ed. Muḥammad Luṭfī al-Ṣabbāgh (Beirut: al-Maktab al-Islāmī, 1996), 8–9.

20 Al-Shāfi'ī, *al-Risāla*, ed. Aḥmad Shākir (Beirut: al-Maktaba al-'Ilmiyya, n.d.), 398–399. It is attributed to Imām Mālik that he allowed narrating from them what had a good or useful meaning but not what was known to be false; Ibn Ḥajar, *Fatḥ al-Bārī*, 6:617.

21 Abū Nu'aym al-Iṣbahānī, *Ḥilyat al-awliyā' wa-ṭabaqāt al-aṣfiyā'*, 10 vols. (Beirut: Dār al-Fikr, and Cairo: Maktabat al-Khānjī, 1996), 9:125.

cannot be sure if their reports are authentic revelation or man-made alteration. One is left with the conclusion reached by al-Kawtharī (d. 1952) a millennium later: that one can accept from them what is affirmed by Islam, reject what is rejected by Islam, and suspend judgment on instances that are unclear.[22] But what if Muslim scholars got this assessment wrong in some instances, like in the case of the Curse of Ham?

Perhaps the scriptures of the People of the Book could be looked at by scholars but not drawn on to formulate law or theology? The Shāfiʿī scholar al-Khaṭṭābī (d. 386/996) explains that what is allowed is that Muslims can narrate reports from the Biblical tradition without *isnāds* to those sources (i.e., Jewish lore from centuries earlier), since those *isnāds* did not exist.[23] A later hadith master, Shams al-Dīn al-Sakhāwī (d. 902/1497), specifies that this is permission to narrate pietistic stories but nothing related to law (*aḥkām*) or matters of belief (ironically, this affirmed what al-Sakhāwī's opponent in a fierce debate over citing Biblical texts, al-Biqāʿī [d. 885/1480], had also reported as the consensus).[24] This was how the Basran Sufi Mālik b. Dīnār (d. ca. 127/745) had used the books of the People of the Book. He would regularly read and cite Jewish writings (*Tawrāh, Zabūr*) to inspire his intensive piety, trust in, and devotion to God.[25]

The position articulated by al-Sakhāwī and others is affirmed by Shams al-Dīn al-Dhahabī (d. 748/1348), Ibn Ḥajar al-ʿAsqalānī, and Ibn Ḥajar al-Haytamī (d. 974/1567) (al-Dhahabī includes an interesting note that one can learn things like medicine and science from the People of the Book too).[26] Muʿāfā b. Zakariyyā (d. 390/1000), a

---

22 Al-Kawtharī, *Maqālāt*, 128.

23 Abū Sulaymān al-Khaṭṭābī, *Maʿālim al-sunan*, 2nd ed., 4 vols. (Beirut: al-Maktaba al-ʿIlmiyya, 1981), 4:187.

24 Al-Sakhāwī, *al-Maqāṣid al-ḥasana*, 192. See Abū al-Ḥasan al-Biqāʿī, *Kitāb al-Aqwāl al-qawīma fī ḥukm al-naql min al-kutub al-qadīma*, in *In Defense of the Bible: A Critical Edition and an Introduction to al-Biqāʿī's Bible Treatise*, by Walid A. Saleh (Leiden: Brill, 2008), 69–70.

25 See al-Iṣbahānī, *Ḥilyat al-awliyāʾ* (1996), 2:358–359, 368, 372, 375, 376, 380.

26 Al-Dhahabī, *Mīzān al-iʿtidāl fī naqd al-rijāl*, ed. ʿAlī Muḥammad al-Bijāwī, 4 vols. (Beirut: Dār al-Maʿrifa, n.d., reprint of ʿĪsā al-Bābī al-Ḥalabī, 1963-4), 3:470; al-Dhahabī, *Siyar aʿlām al-nubalāʾ*, 3rd ed., ed. Shuʿayb al-Arnāʾūṭ, 25

jurist from the Jarīrī school of law, said that the aim was to prevent people who knew stories about the miracles and wonders of Jewish history from being afraid to share them, so that such wisdom would not be lost.[27] The opinion of al-Muhallab (d. 425/1044) (affirmed by Ibn Baṭṭāl and Ibn Ḥajar) is that Muslims are prohibited from seeking information from the People of the Book on matters on which our law has a ruling, for "our law is sufficient." As for matters on which there is no text (*naṣṣ*) from our law, "then one can examine it and use it as an indication (*istidlāl*), as long as the law [of Islam] supports it (*mā yaqūmu al-shar' minhu*)."[28]

But, again, this may be kicking the can down the road. There is the risk of error in determining if our law supports a story or not. And there is some inherent danger in relying on previous scriptures. Such a warning comes from Imām al-Māturīdī (d. 333/944). Discussing the Qur'ānic story of Moses' encounter with Pharaoh's sorcerers, al-Māturīdī cautions against accepting additional, innocuous-seeming information from the People of the Book. This was because of "the additions and subtractions that enter into [this material], which give unbelievers a point to make in impugning the message of God's Prophet (ṣ)," since the need for such details suggests something is wanting in Islam's revelation.[29]

These problems were moot if Muslims turned not to the books of previous religious communities to flesh out Qur'ānic stories but rather to the teachings of the Prophet (ṣ) himself. Many hadiths are stories the Prophet (ṣ) told about earlier messengers. Provided such

---

vols. (Beirut: Mu'assasat al-Risāla, 1992–1998), 3:86; Ibn Ḥajar al-Haytamī, *al-Fatāwā al-ḥadīthiyya*, ed. Muḥammad 'Abd al-Raḥmān al-Mar'ashlī (Beirut: Dār Iḥyā' al-Turāth al-'Arabī, 1998), 15–16. See also al-Ṭaḥāwī, *Sharḥ mushkil al-āthār*, 9:267–272. 'Abd Allāh b. al-Ṣiddīq al-Ghumārī (d. 1993) holds that questions of faith, *uṣūl*, and *furū'* cannot be taken from the *ahl al-kitāb*, but preaching (*mawā'iẓ*), stories, *targhīb*, and *ādāb* can; al-Ghumārī, *Afḍal maqūl fī manāqib afḍal rasūl* (Cairo: Maktabat al-Qāhira, 2005), 79.

27 M.J. Kister, "'Ḥaddithū 'an banī isrā'īl wa lā ḥaraj': A Study of an Early Tradition," *Israel Oriental Studies* 2 (1979).

28 Ibn Baṭṭāl, *Sharḥ Ṣaḥīḥ al-Bukhārī*, 10:391; cf. Ibn Ḥajar, *Fatḥ al-Bārī*, 13:412.

29 Abū al-Manṣūr al-Māturīdī, *Ta'wīlāt al-Qur'ān*, ed. Bekir Topaloglu et al., 17 vols. (Istanbul: Dār al-Mīzān, 2006), 10:296.

hadiths could be authenticated, they would represent sound sources for details about Qur'ānic stories and figures from earlier times. One example is the Prophet's lengthy and detailed account of the journey of Moses and the sage.[30]

Here, of course, we find ourself before the problem astutely identified by some Islamic modernists: Muslim scholars were often not as assiduous as they should have been in weeding out authentic hadiths from *isrāʾīliyyāt* that had been mixed up with hadiths and wrongly attributed to the Prophet (ṣ). This could have happened through outright forgery, but it was also accidental. Muslim b. Ḥajjāj (d. 261/875) recounts how, even in the early decades of Islam, scholars could mix up stories circulating from Jewish or Christian converts to Islam with the Prophet's (ṣ) words. The Successor Busr b. Saʿīd (d. 100/719) recalled that he would attend the lectures of Abū Hurayra, who would narrate both reports from the Prophet (ṣ) and stories from the early Jewish convert to Islam, Kaʿb al-Aḥbār. Busr later overheard some in the audience mixing up what were the words of the Prophet (ṣ) and what were Kaʿb's stories.[31] Such a mix-up occurred in a report about the importance of Friday (that it was the day Adam was created and that the Day of Judgment will occur), according to Ibn Khuzayma. He describes how he had collected the various narrations of this report, some tracing it back through Abū Hurayra to the Prophet (ṣ) and some through Abū Hurayra to Kaʿb al-Aḥbār, with Ibn Khuzayma favoring the second version as the more accurate.[32]

Of course, such mix-ups could (and did) occur with all sorts of reports and hadiths, not just *isrāʾīliyyāt*. But what raised the risk of error significantly with *isrāʾīliyyāt* was that Muslim scholars tended to be laxer when dealing with reports describing the End Times and Qur'ānic stories, since they were not considered as important as

30 Ibn Ḥajar says that Abū Jaʿfar b. al-Munādī (d. 272/886) compiled a book on reports regarding the mystical sage al-Khaḍir and that he concluded that there was no *isrāʾīliyyāt* material on that subject that could be relied upon; Ibn Ḥajar, *Fatḥ al-Bārī*, 8:53; Ibn Ḥajar al-ʿAsqalānī, *al-Zahr al-naḍir fī ḥāl al-Khaḍir*, ed. Ṣalāḥ Maqbūl Aḥmad (New Delhi: Majmaʿ al-Buḥūth al-Islāmiyya, 1988), 57.

31 Muslim b. Ḥajjāj, *Kitāb al-Tamyīz*, 175.

32 Ibn Khuzayma, *Ṣaḥīḥ Ibn Khuzayma*, ed. Muḥammad Muṣṭafā al-Aʿẓamī, 4 vols. (Beirut: al-Maktab al-Islāmī, [1970]), 3:115.

matters of law and theology.[33] And these areas of laxity were precisely the topics on which *isrāʾīliyyāt* tended to touch. Hence the problem bemoaned by Ibn al-Jawzī eight centuries ago and by the late Egyptian scholar Shaykh Muḥammad al-Ghazālī (d. 1996) just three decades ago: that *tafsīr* books are full of unreliable hadiths and *isrāʾīliyyāt.*[34]

Of course, many Muslim scholars were extremely vigilant about identifying *isrāʾīliyyāt.* In one report, Ibn ʿAbbās is told that Kaʿb al-Aḥbār has been teaching that on the Day of Judgment the sun and the moon will be brought "like two bulls for the slaughter and cast into the Fire." Ibn ʿAbbās objects vehemently, exclaiming that the sun and moon are slaves of God that have done nothing deserving punishment and that this was nothing more than "Jewish lore" (*yahūdiyya*) that Kaʿb was trying to insinuate into Islam.[35] Sometimes they might be overly vigilant. Burhān al-Dīn al-Bayjūrī (d. 1860) resolves the conflict between the theological tenet that angels have no capacity to disobey God and the Qurʾānic story of Hārūt and Mārūt descending to earth and engaging in evil by contending that Hārūt and Mārūt

33 Jonathan A.C. Brown, "Even If It's Not True It's True: Using Unreliable Ḥadīths in Sunni Islam," *Islamic Law and Society* 18 (2011).

34 Ibn al-Jawzī, *al-Quṣṣāṣ*, 103; Muḥammad al-Ghazālī, *Turāthunā al-fikrī fī mīzān al-sharʿ wa-l-ʿaql*, 8th ed. (Cairo: Dār al-Shurūq, 2003), 126–128.

35 Al-Ṭabarī, *Tārīkh al-rusul wa-l-mulūk*, 6 vols. (Beirut: Dār al-Kutub al-ʿIlmiyya, 2003), 1:47. The issue is more complicated than the opinion attributed to Ibn ʿAbbās suggests. There are several versions of this hadith, some stronger and less objectionable than others. Al-Bukhārī includes the hadith "The sun and the moon are rolled up on the Day of Judgment," which does not differ much in meaning from Q. 81:1; *Ṣaḥīḥ al-Bukhārī*: kitāb badʾ al-khalq, bāb ṣifat al-shams wa-l-qamar. The problematic versions of the hadith are "The sun and the moon are two bulls rolled up in the Fire on the Day of Judgment" (narrated via Abū Hurayra in Abū al-Qāsim Tammām, *Fawāʾid*, ed. Ḥamdī ʿAbd al-Majīd al-Salafī, 2 vols. [Riyadh: Maktabat al-Rushd, 2002], 2:204) and "The sun and the moon are two bulls slaughtered in the Fire" (narrated via Anas b. Mālik in Abū Dāwūd al-Ṭayālisī, *Musnad*, ed. Muḥammad ʿAbd al-Muḥsin al-Turkī, 4 vols. [Cairo: Dār Hajr, 1999], 3:574). Al-Khaṭṭābī, however, argues that Ibn ʿAbbās's objection was misplaced. Being placed in the Fire was not a punishment for the sun and the moon but rather them serving to rebuke those people who had worshipped them and would then be reminded of that in the Fire; al-Khaṭṭābī, *Aʿlām al-sunan*, ed. Muḥammad Saʿd Āl Suʿūd, 4 vols. (Mecca: Jāmiʿat Umm al-Qurā, 1988), 2:1476–1477.

had never been angels (*malak*) but rather just two righteous men (one Qur'ānic reading vowels the word *malik*, or king). Reports that they were angels, he argues, are not reliable and nothing more than *isrā'īliyyāt* wrongly accepted by Muslim historians.[36]

Sometimes the can, long kicked, finally got away from Muslim scholars. In some instances, they did not grasp what was entailed theologically or legally in the *isrā'īliyyāt* they accepted or did not notice that they clearly violated a tenet of Islamic law or theology. To quote al-Shāfi'ī, sometimes one might be gathering wood and not realize there was a viper in the bundle.[37] For example, discussions of the story of David seeking God's forgiveness (Q. 38:21–25) in most *tafsīr* books include reports about the Biblical episode of David and Bathsheba (2 Samuel 11), in which King David sees the married Bathsheba bathing, becomes infatuated with her, then has her husband sent off to die in a war so that he can marry her. In *tafsīr* works these reports are narrated from early authorities like Ismā'īl al-Suddī (d. 127/745) and the storyteller Wahb b. Munabbih, wrongly attributed to prominent early scholars like Ibn 'Abbās and al-Ḥasan al-Baṣrī (d. 110/728), and even wrongly attributed to the Prophet (ṣ).[38] Some *tafsīr* scholars like al-Qurṭubī (d. 671/1272), al-Dāwūdī (d. 467/1075), and Ibn al-Jawzī remark with outrage that such reports are not reliably transmitted and that their meaning is unacceptable.[39] Ibn Kathīr and al-Biqā'ī even state explicitly that the David and Bathsheba story is a clear case of *isrā'īliyyāt*.[40] The Cordoban Qāḍī 'Iyāḍ (d. 544/1149) tells his readers to ignore such

36 Al-Bayjūrī, *Ḥāshiyat al-Imām al-Bayjūrī*, 222. This was also suggested by Abū Bakr al-Bāqillānī, *Kitāb al-Bayān*, 80.

37 Ibn 'Adī, *al-Kāmil fī ḍu'afā' al-rijāl*, 7 vols. (Beirut: Dār al-Fikr, 1985), 1:124.

38 See *Tafsīr al-Ṭabarī*, for example, on Q. 38:21–25.

39 Ibn al-Jawzī, *al-Quṣṣāṣ*, 103; Qāḍī 'Iyāḍ, *Kitāb al-Shifā'*, ed. 'Abduh 'Alī Kūshak (Manama: Maktabat Niẓām Ya'qūbī, 2015), 694–695, and see next note.

40 Al-Qurṭubī, *al-Jāmi' li-aḥkām al-Qur'ān*, ed. Muḥammad Ibrāhīm al-Ḥifnāwī and Maḥmūd Ḥāmid 'Uthmān, 20 vols. in 10 (Cairo: Dār al-Ḥadīth, 1994), 8:142, 155; Ibn al-Jawzī, *Zād al-masīr*, ed. 'Abd al-Razzāq Mahdī, 4 vols. (Beirut: Dār al-Kitāb al-'Arabī, 2002), 3:566; Abū al-Ḥasan al-Biqā'ī, *Naẓm al-durar*, 22 vols. (Cairo: Dār al-Kitāb al-Islāmī, 1984), 16:361–362.

stories that have been borrowed from the People of the Book and have no sound narrations.[41]

Unfortunately, the list of *tafsīr* works that include the David and Bathsheba story is not short, and the list of authors on it includes many luminaries. Thank God for the vigilance of scholars like al-Qurṭubī and Ibn al-Jawzī, and thank God for a tradition that continues to produce scholars like Muntasir Zaman, *ḥafiẓahu Allāh.*

41 Qāḍī ʿIyāḍ, op. cit.

APPENDIX I

In chapter five, we analyzed the various routes of hadith describing Prophet Adam's height as sixty cubits and humankind's gradual decrease in height thereafter. We identified four primary hadith bundles, each comprising routes that are either inclusive or exclusive of these two clauses. To provide some context for our analysis, we reproduce the Arabic wording of the examined hadith below. Nearly seventy routes of transmission were analyzed, so it is not feasible to reproduce the wording of each route. We present at least two routes from each hadith bundle, highlighting one that is inclusive and another that is exclusive of the height description. A basic reference is provided after each hadith. For detailed citations, refer to chapter five.

Number	Arabic Wording
	عن أبي زرعة، عن أبي هريرة، عن النَّبيّ ﷺ، قال: «إنّ أوّل زمرة يدخلون الجنّة على صورة القمر ليلة البدر، ثمّ الّذين يلونهم على أشدّ كوكب دُرّيّ في السّماء إضاءة، لا يبولون، ولا يتغوّطون، ولا يتفلون ولا يمتخطون، أمشاطهم الذّهب، وَرَشْحهم المسك، ومَجامِرهم الألوّة: الأَنجُوج، عود الطيب وأزواجهم الحور العين، على خلق رجل واحد، **على صورة أبيهم آدم، ستّون ذراعاً في السّماء**» أخرجه البخاري في صحيحه (٧٢٣٣).
Hadith 1	عن عبد الرّحمن بن أبي عَمْرة، عن أبي هريرة، عن النَّبيّ ﷺ، قال: «أوّل زمرة تدخل الجنّة على صورة القمر ليلة البدر، والّذين على آثارهم كأحسن كوكب دُرّيّ في السّماء إضاءةً، قلوبهم على قلب رجل واحد، لا تَباغض بينهم ولا تَحاسد، لكل امرئ زوجتان من الحور العين، يُرى مُخّ سُوقهن من وَراء العَظم واللّحم» أخرجه البخاري في صحيحه (٤٥٢٣).
	عن أبي سعيد الخدري، عن النَّبيّ ﷺ، قال: «أوّل زمرة يدخلون الجنّة يوم القيامة ضوء وجوههم على مثل ضوء القمر ليلة البدر، والزّمرة الثّانية على مثل أحسن كوكب دُرّيّ في السّماء، لكلّ رجل منهم زوجتان على كلّ زوجة سبعون حُلّة يُرى مُخّ ساقها من وَرائها» أخرجه الترمذي في جامعه (٥٣٥٢).

Number	Arabic Wording
Hadith 2	عن همّام بن منبّه، عن أبي هريرة، عن النّبيّ ﷺ، قال: «خلق الله آدم وطوله **ستّون ذراعاً**، ثمّ قال: اذهب فسلّم على أولئك من الملائكة، فاستمع ما يحيّونك، تحيّتك وتحيّة ذرّيّتك، فقال السّلام عليكم، فقالوا: السّلام عليك ورحمة الله، فزادوه: ورحمة الله، فكلّ من يدخل الجنّة على صورة آدم، **فلم يزل الخلق ينقص حتّى الآن**» أخرجه البخاري في صحيحه (٦٢٣٣). عن سعيد بن أبي سعيد، عن أبي هريرة، عن النّبيّ ﷺ، قال: «لمّا خلق الله آدم ونفخ فيه الرّوح عطس، فحمد ربّه بإذن الله له فقال: الحمد لله، فقال له ربّه: رحمك ربّك يا آدم، اذهب إلى أولئك الملأ، وملأ منهم جلوس، فقل: السّلام عليكم، فقالوا: سلام عليك ورحمة الله، ثمّ رجع إلى ربّه فقال: هذه تحيّتك وتحيّة ذرّيّتك بينهم» أخرجه النسائي في السنن الكبرى (٥٧٩٩).
Hadith 3	عن سعيد بن المسيّب، عن أبي هريرة، عن النّبيّ ﷺ، قال: «يَدخل أهل الجنّةِ الجنّةَ جردا، مردا، بيضا، جعادا، مكحّلين، أبناء ثلاث وثلاثين، على خلق آدم، **ستّون ذراعاً في عرض سبع أذرع**» أخرجه أحمد في مسنده (٣٣٩٧) عن معاذ بن جبل، عن النّبيّ ﷺ، قال: «يَدخل أهل الجنّة الجنّةَ جردا، مردا، مكحّلين، بني ثلاثين أو ثلاث وثلاثين» أخرجه أحمد في مسنده (٦٠١٢٢). عن أنس بن مالك، عن النّبيّ ﷺ، قال: «يَدخل أهل الجنّةِ الجنّةَ جردا، مردا، مكحّلين» أخرجه الطبراني في المعجم الصغير (٤٦١١).

Number	Arabic Wording
	عن الحسن، أُبيّ بن كعب، عن النّبيّ ﷺ، قال: «إنّ أباكم آدم كان طوالا **كالنّخلة السّحوق، ستّين ذراعاً**، كثير الشّعر، مُوارى العورة، وأنه لمّا أصاب الخطيئة بدَت له سَوءته فخرج هارباً في الجنّة فتلقاه شجرة، فأخذت بناصيته، وناداه ربّه: أفِراراً منّي يا آدم؟ قال: لا واللهِ يا ربّ، ولكن حياءً منك ممّا قد جَنيتُ» أخرجه الطبري في تاريخه (١:١٦٠).
Hadith 4	عن عُتَيّ، عن أُبيّ بن كعب، عن النّبيّ ﷺ، قال: «إنّ آدم كان رجلا طوالا كأنّه نخلة سَحوق، كثير شعر الرّأس، فلمّا ركب الخطيئة بدَت له عَورته وكان لا يَراها قبل ذلك، فانطلق هارباً في الجنّة فتعلّقت به شجرة، فقال لها: أرسليني، فقالت: لستُ بمُرسِلتك، قال: وناداه ربّه: أمِنّي تَفِرّ؟ قال: ربّ إنّي استَحييتُ» أخرجه ابن سعد في الطبقات الكبرى (١:٣١).

APPENDIX II

The following list summarizes the findings of our analysis in chapter five, enumerating the number of inclusive and exclusive routes. The focus is to represent the number of routes; the status of many of these hadith is discussed in chapter five. This list does not include Successor reports (e.g., Qatāda's inclusive statement under Hadith 3) or fabricated hadith (e.g., Jābir's exclusive version of Hadith 3). Under the "inclusion" column, the clause regarding the gradual decrease of human height only features under Hadith 2; it is noted with an asterisk. It goes without saying that the data presented here is based on the routes that I was able to locate at the time of conducting the analysis. Further research may potentially modify these findings.

Number	Transmitter	Inclusion (18)	Exclusion (40)
Hadith 1	Abū Hurayra	2	9
	Abū Saʿīd al-Khudrī		1
	Jābir b. ʿAbd Allāh		1
	Anas b. Mālik		1
Hadith 2	Abū Hurayra	3*	6
	ʿAbd Allāh b. ʿUmar (*mawqūf*)	1*	
	ʿAbd Allāh b. Salām (*mawqūf*)		1

Number	Transmitter	Inclusion (18)	Exclusion (40)
Hadith 3	Abū Hurayra via Ḥammād	6	2
	Abū Hurayra via ʿĀmir		1
	Muʿādh b. Jabal		1
	Anas via al-Awzāʿī	1	3
	Anas via Yaḥyā		1
	Anas via Abū ʿĀtika		1
	Miqdām b. Maʿdīkarib		1
	Miqdād b. al-Aswad		1
	Abū Lubāba		1
	[ʿAbd Allāh] b. Busr		1
	Wāthila b. al-Asqaʿ		1
	ʿAbd Allāh b. ʿAbbās		1
	ʿAbd Allāh b. Masʿūd		1
	Ḥasan al-Baṣrī	1	
Hadith 4	Ubayy via al-Ḥasan	2	1
	Ubayy via ʿUtayy		1
	Ubayy (*mawqūf*)		1
	Anas b. Mālik		1
Miscellanea	Abū Hurayra	1	
	Ubayy b. Kaʿb (*mawqūf*)	1	
	Abū Hurayra		1

BIBLIOGRAPHY

Abd-Allah, Umar F. *Mālik and Medina: Islamic Legal Reasoning in the Formative Period.* Leiden: Brill, 2013.

ʿAbd al-Jabbār, Abū al-Ḥasan. *al-Mughnī fī abwāb al-tawḥīd wa-l-ʿadl.* Edited by Amīn al-Khūlī. Cairo: al-Dār al-Miṣriyya li-l-Taʾlīf wa-l-Tarjama, [195-?]–[196-?].

Abū Dāwūd al-Sijistānī. *Risālat al-Imām Abī Dāwūd al-Sijistānī ilā Ahl Makka.* In *Thalāth rasāʾil fī ʿilm muṣṭalaḥ al-ḥadīth,* by ʿAbd al-Faṭṭāh Abū Ghudda, 29–54. Beirut: Dār al-Bashāʾir al-Islāmiyya, 2005.

Abū Dāwūd al-Ṭayālisī. *Musnad.* Edited by Muḥammad ʿAbd al-Muḥsin al-Turkī. 4 vols. Cairo: Dār Hajr, 1419/1999.

Abū Ḥudhayfa, Nabīl. *Anīs al-sārī fī taḥqīq wa-takhrīj al-aḥādīth allatī dhakarahā al-Ḥāfiẓ Ibn Ḥajar al-ʿAsqalānī fī Fatḥ al-Bārī.* Beirut: Muʾassasat al-Rayyān, 2005.

Abū Rayya, Maḥmūd. *Aḍwāʾ ʿalā al-Sunna al-Muḥammadiyya.* Cairo: Dār al-Taʾlīf, 1958.

Abū Rayya, Maḥmūd. "Kaʿb al-Aḥbār huwa al-ṣahyūnī al-awwal." *al-Risāla wa-l-Riwāya* 665 (1946): 360–362.

Abū Shahba, Muḥammad. *Difāʿ ʿan al-sunna wa-radd shubah al-mustashriqīn wa-l-kuttāb al-muʿāṣirīn.* Cairo: Maktabat al-Sunna, 1989.

Abū Shahba, Muḥammad. *al-Isrāʾīliyyāt wa-l-mawḍūʿāt fī kutub al-tafsīr.* Cairo: Maktabat al-Sunna, 1988.

Adang, Camilla. *Muslim Writers on Judaism and the Hebrew Bible.* Leiden: Brill, 1996.

Akbarābādī, Saʿīd. *ʿUthmān Dhū al-nūrayn.* Mumbai: Maktabat al-Ḥaqq, 2013.

al-ʿAlāʾī, Salāḥ al-Dīn. *al-Mukhtaliṭīn.* Cairo: Maktabat al-Khānjī, 1996.

al-Albānī, Nāṣir al-Dīn. *Silsilat al-aḥādīth al-ṣaḥīḥa wa-shayʾ min fiqhihā wa-fawāʾidihā.* Riyadh: Maktabat al-Maʿārif, 1417/1996.

al-Ālūsī, Maḥmūd Shukrī. *Rūḥ al-maʿānī fī tafsīr al-Qurʾān al-karīm wa-l-sabʿ al-mathānī.* Beirut: Dār al-Kutub al-ʿIlmiyya, 1994.

Angel, Lawrence. "The Bases of Paleodemography." *American Journal of Physical Anthropology* 30, no. 3 (1969): 427–437.

al-Aṣfahānī, Abū Nuʿaym. *Ḥilyat al-awliyāʾ wa-ṭabaqāt al-aṣfiyāʾ*. Beirut: Dār al-Kutub al-ʿIlmiyya, 1988.

al-Aṣfahānī, Abū Nuʿaym. *Ṣifat al-janna*. Damascus: Dār al-Maʾmūn, 1995.

al-ʿĀṣimī, ʿAbd al-Malik. *Samṭ al-nujūm al-ʿawālī fī anbāʾ al-awāʾil wa-l-tawālī*. Beirut: Dār al-Kutub al-ʿIlmiyya, 1998.

Auerbach, Benjamin. Review of *The Ancient Giants Who Ruled America: The Missing Skeletons and the Great Smithsonian Cover-Up*, by Richard J. Dewhurst. *American Antiquity* 80, no. 3 (2013): 628–629.

al-ʿAwnī, Ḥātim b. ʿĀrif. *al-Mursal al-khafī wa-ʿilāqatuhu bi-l-tadlīs*. Riyadh: Dār al-Hijra, 1997.

al-ʿAwnī, Ḥātim b. ʿĀrif. *Sharḥ al-ḥadīth al-nabawī: dirāsa fī tārīkh li-l-ʿilm wa-l-taʾṣīl lahu wa-taqwīm al-muṣannafāt fīhi wa-l-tadrīb ʿalayhi*. Beirut: Markaz al-Namāʾ, 2021.

al-ʿAwnī, Ḥātim b. ʿĀrif. *al-Tamarrud ʿalā al-ʿilm (al-qadīm) bi-daʿwā ʿilm (al-ḥadīth)*. al-Madīna News. June 13, 2013. https://www.al-madina.com/article/236521.

ʿAwwāma, Muḥammad. *Ḥadhf ṭaraf min al-ḥadīth al-wāḥid ikhtiṣāran aw iʿlālan*. Jeddah: Dār al-Minhāj, 2017.

ʿAwwāma, Muḥammad. *Hal fī ḥadīth « khalaqa Allāh al-turba yawm al-sabt » ishkāl?* Jeddah: Dār al-Minhāj, 2016.

al-ʿAynī, Badr al-Dīn. *ʿUmdat al-qārī fī sharḥ Ṣaḥīḥ al-Bukhārī*. Beirut: Dār Iḥyāʾ al-Turāth al-ʿArabī, n.d.

al-Aʿẓamī, Ḍiyāʾ al-Raḥmān. "Abū Hurayra fī ḍawʾ marwiyyātihi bi-shawāhidihā." MA thesis, Jāmiʿat al-Malik ʿAbd al-ʿAzīz (Jeddah), 1973.

ʿAzzāz, ʿAbd al-Raḥmān. "al-Tawaqquf fī al-masāʾil al-uṣūliyya fī dalālāt al-alfāẓ." MA thesis, Jāmiʿat al-Imām Muḥammad b. Saʿūd al-Islāmiyya (Riyadh), 2005.

Bāʿalawī, Muḥammad b. Abī Bakr. *al-Mushriʿ al-rawī fī manāqib al-sāda al-kirām Āl Abī ʿAlawī*. Cairo: al-Maṭbaʿa al-ʿĀmira, 1319 AH.

al-Baghawī, al-Ḥusayn. *Sharḥ al-sunna*. Edited by Shuʿayb al-Arnāʾūṭ and Zuhayr al-Shāwīsh. 15 vols. Beirut: al-Maktab al-Islāmī, 1983.

Baḥshal, Abū al-Ḥasan. *Tārīkh Wāsiṭ*. Beirut: ʿĀlam al-Kutub, 1406 AH.

al-Bājī, Abū al-Walīd. *Taḥqīq al-madhhab*. Riyadh: ʿĀlam al-Kutub, 1983.

al-Bakrī, Ḥamza Muḥammad. *Taʿaddud al-ḥāditha fī riwāyāt al-ḥadīth al-nabawī: dirāsa taʾṣīliyya naqdiyya*. Amman: Arwiqa, 2013.

Bannūrī, Muḥammad Yūsuf. *Maʿārif al-sunan sharḥ Jāmiʿ al-Tirmidhī*. Karachi: H.M. Saʿīd, 1992.

al-Bāqillānī, Abū Bakr. *Kitāb al-Bayān ʿan al-farq bayna al-muʿjizāt wa-l-karāmāt*. Edited by Richard J. McCarthy. Beirut: al-Maktaba al-Sharqiyya/Librairie Orientale, 1958.

Bāshanfar, Saʿīd. *Manhaj al-Imām al-Bukhārī fī ʿarḍ al-ḥadīth al-maʿlūl fī al-Jāmiʿ al-ṣaḥīḥ*. Beirut: Dār Ibn Ḥazm, 2016.

al-Bayhaqī, Abū Bakr. *al-Baʿth wa-l-nushūr*. Riyadh: Maktabat Dār al-Ḥijāz, 2013.

al-Bayhaqī, Abū Bakr. *Kitāb al-Asmāʾ wa-l-ṣifāt*. Edited by Muḥammad Zāhid al-Kawtharī. Cairo: al-Maktaba al-Azhariyya, n.d.

al-Bayhaqī, Abū Bakr. *al-Sunan al-kubrā*. Beirut: Dār al-Kutub al-ʿIlmiyya, 2003.

al-Bayjūrī, Burhān al-Dīn. *Ḥāshiyat al-Imām al-Bayjūrī ʿalā Jawharat al-tawḥīd*. Cairo: Dār al-Salām, 2006.

al-Bazzār, Abū Bakr. *al-Musnad*. Medina: Maktabat al-ʿUlūm wa-l-Ḥikma, 2009.

al-Biqāʿī, Abū al-Ḥasan. *Kitāb al-Aqwāl al-qawīma fī ḥukm al-naql min al-kutub al-qadīma*. In *In Defense of the Bible: A Critical Edition and an Introduction to al-Biqāʿī's Bible Treatise*, by Walid A. Saleh. Leiden: Brill, 2008.

al-Biqāʿī, Abū al-Ḥasan. *Naẓm al-durar*. 22 vols. Cairo: Dār al-Kitāb al-Islāmī, 1404/1984.

Bīrī Zādah, Muḥammad. *al-Fatḥ al-Raḥmānī sharḥ Muwaṭṭaʾ al-Imām Muḥammad b. al-Ḥasan al-Shaybānī*. Istanbul: Yūsuf Āghā Library, manuscript no. 338.

al-Bisṭāmī, Muḥammad. "al-Muṭahhar al-Maqdisī wa-manhajuhu al-tārīkhī fī *Kitāb al-Badʾ wa-l-tārīkh*." PhD dissertation, al-Azhar University, 2008.

Blecher, Joel. *Said the Prophet of God: Hadith Commentary across a Millennium*. Oakland: University of California Press, 2018.

Bogaard, Amy. "Communities." In *The Cambridge World History*, vol. 2, *A World with Agriculture, 12,000 BCE–500 CE*, edited by Graeme Baker and Candice Goucher, 124–160. Cambridge: Cambridge University Press, 2015.

Bondeson, Jan. *A Cabinet of Medical Curiosities*. New York: Cornell University Press, 1997.

Bonner, John. *Why Size Matters: From Bacteria to Blue Whales*. Princeton: Princeton University Press, 2006.

Brown, Daniel W. *Rethinking Tradition in Modern Islamic Thought*. Cambridge: Cambridge University Press, 1996.

Brown, Jonathan A.C. *The Canonization of al-Bukhārī and Muslim: The Formation and Function of the Sunnī Ḥadīth Canon*. Leiden: Brill, 2007.

Brown, Jonathan A.C. "The Canonization of Ibn Mâjah: Authenticity vs. Utility in the Formation of the Sunni Ḥadîth Canon." *Revue des mondes musulmans et de la Méditerranée* 129 (2011): 169–181.

Brown, Jonathan A.C. "Did the Prophet Say It or Not? The Literal, Historical and Effective Truth of *Ḥadīths*." *Journal of the American Oriental Society* 129, no. 2 (2009): 259–285.

Brown, Jonathan A.C. "Even If It's Not True It's True: Using Unreliable Ḥadīths in Sunni Islam." *Islamic Law and Society* 18 (2011): 1–52.

Brown, Jonathan A.C. "Faithful Dissenters: Sunni Skepticism about the Miracles of Saints." *Journal of Sufi Studies* 1, no. 2 (2012): 123–168.

Brown, Jonathan A.C. *Hadith: Muhammad's Legacy in the Medival and Modern World*. Oxford: Oneworld Publications, 2009.

Brown, Jonathan A.C. "How We Know Early Ḥadīth Critics Did *Matn* Criticism and Why It's So Hard to Find." *Islamic Law and Society* 15 (2008): 143–184.

Brown, Jonathan A.C. *Misquoting Muhammad: The Challenge and Choices of Interpreting the Prophet's Legacy*. Oxford: Oneworld, 2014.

Brown, Jonathan A.C. "The Rules of *Matn* Criticism: There Are No Rules." *Islamic Law and Society* 19 (2012): 356–396.

Brown, Jonathan A.C. *Slavery and Islam*. London: Oneworld, 2019.

Brusatte, Stephen. *Dinosaur Paleobiology*. Hoboken: Wiley-Blackwell, 2012.

al-Bukhārī, 'Alā' al-Dīn. *Kashf al-asrār 'an Uṣūl Fakhr al-Islām al-Bazdawī*. Beirut: Dār al-Kitāb al-Islāmī, n.d.

al-Bukhārī, Muḥammad b. Ismā'īl. *al-Jāmi' al-musnad al-ṣaḥīḥ al-mukhtaṣar min umūr Rasūl Allāh wa-sunanihi wa-ayyāmihi*. Edited by Zuhayr Nāṣir. 9 vols. Beirut: Dār Ṭawq al-Najāh, 1422 AH.

al-Bukhārī, Muḥammad b. Ismā'īl. *al-Tārīkh al-kabīr*. Hyderabad: Dā'irat al-Ma'ārif al-'Uthmāniyya, n.d.

al-Bustī, Ibn Ḥibbān. *Ṣaḥīḥ Ibn Ḥibbān*. Beirut: Mu'assasat al-Risāla, 1993.

Conrad, Lawrence. "Seven and the Tasbi': On the Implications of Numerical Symbolism for the Study of Medieval Islamic History." *Journal of the Economic and Social History of the Orient* 31 (1988): 42–73.

Dajani, Rana. "Evolution and Islam: Is There a Contradiction?" In *Islam and Science: Muslim Responses to Science's Big Questions*, edited by Usama Hasan and Athar Osama, 142–151. London: Ihsanoglu Task Force, 2016.

Dānāpūrī, 'Abd al-Ra'ūf. *Aṣaḥḥ al-siyar fī hady khayr al-bashar*. Karachi: Aṣaḥḥ al-Maṭābi', n.d.

al-Dāraquṭnī, 'Alī b. 'Umar. *al-'Ilal al-wārida fī al-aḥādīth al-nabawiyya*. Riyadh: Dār Ṭayba, 1985.

al-Dāraquṭnī, ʿAlī b. ʿUmar. *al-Muʾtalif wa-l-mukhtalif*. Beirut: Dār al-Gharb al-Islāmī, 1986.

al-Dāraquṭnī, ʿAlī b. ʿUmar. *Sunan al-Dāraquṭnī*. Beirut: Muʾassasat al-Risāla, 2004.

al-Dārimī, ʿAbd Allāh. *Sunan al-Dārimī*. Riyadh: Dār al-Mughnī, 2000.

al-Dhahabī, Shams al-Dīn. *Mīzān al-iʿtidāl fī naqd al-rijāl*. Edited by ʿAlī Muḥammad al-Bijāwī. 4 vols. Beirut: Dār al-Maʿrifa, n.d. Reprint of ʿĪsā al-Bābī al-Ḥalabī, 1963-4.

al-Dhahabī, Shams al-Dīn. *Siyar aʿlām al-nubalāʾ*. 3rd ed. Edited by Shuʿayb al-Arnāʾūṭ. 25 vols. Beirut: Muʾassasat al-Risāla, 1992–1998.

Dihlawī, Shāh Walī Allāh. *Ḥujjat Allāh al-bāligha*. Beirut: Dār al-Jīl, 2005.

Dihlawī, Shāh Walī Allāh. *al-Tafhīmāt al-ilāhiyya*. Bijnor: Madīna Barqī Press, 1936.

Dihlawī, Shāh Walī Allāh. *Taʾwīl al-aḥādīth fī rumūz qaṣas al-anbiyāʾ*. Hyderabad: Akādīmiyyat al-Shāh Walī Allāh, 1966.

Donnelly, Deidre, and Patrick Morrison. "Hereditary Gigantism: The Biblical Giant Goliath and His Brothers." *Ulster Medical Journal* 83, no. 2 (2014): 86–88.

Duyar, İzzet, and Barış Özener. "Evolution of Human Body Height and Its Implications in Ergonomics." *Gaziantep Üniversitesi Sosyal Bilimler Dergisi* 1 (2009): 63–75.

Efendī, Abū al-Suʿūd. *Irshād al-ʿaql al-salīm ilā mazāyā al-Kitāb al-Karīm*. Beirut: Dār Iḥyāʾ al-Turāth al-ʿArabī, n.d.

El Shamsy, Ahmed. *Rediscovering the Islamic Classics: How Editors and Print Culture Transformed an Intellectual Tradition*. Princeton: Princeton University Press, 2020.

El Shamsy, Ahmed. "The Ur-*Muwaṭṭaʾ* and Its Recensions." *Islamic Law and Society* (published online ahead of print, 2021): 1–30.

El-Tobgui, Carl Sharif. *Ibn Taymiyya on Reason and Revelation: A Study of* Darʾ taʿāruḍ al-ʿaql wa-l-naql. Leiden: Brill, 2020.

Falāta, ʿUmar. *al-Waḍʿ fī al-ḥadīth*. Damascus: Maktabat al-Ghazālī, 1981.

al-Faramāwī, ʿUmar Muḥammad. "Min masālik al-muḥaddithīn wa-l-uṣūliyyīn fī al-taʿāmul maʿa mukhtalif al-ḥadīth wa-mushkilihi." *Majallat Dār al-Iftāʾ al-Miṣriyya* 19, no. 1 (2014): 52–119.

al-Fasawī, Yaʿqūb b. Sufyān. *al-Maʿrifa wa-l-tārīkh*. Beirut: Muʾassasat al-Risāla, 1981.

al-Fayyūmī, Aḥmad. *al-Qawl al-tāmm fī bayān aṭwār sayyidinā Ādam 'alayhi al-ṣalāh wa-afḍal al-salām*. Riyadh: King Abdul Aziz Library, manuscript no. 684.

Feder, Kenneth. *Frauds, Myths, and Mysteries: Science and Pseudoscience in Archaeology*. New York: McGraw Hill, 2013.

al-Firyābī, Ja'far. *Kitāb al-Qadar*. Riyadh: Aḍwā' al-Salaf, 1997.

Gallagher, Andrew. "Stature, Body Mass, and Brain Size: A Two-Million-Year Odyssey." *Economics and Human Biology* 11 (2013): 551–562.

Ghani, Usman. "'Abū Hurayrah' a Narrator of Ḥadīth Revisited: An Examination into the Dichotomous Representations of an Important Figure with Special Reference to Classical Islamic Modes of Criticism." PhD dissertation, University of Exeter, 2011.

al-Ghazālī, Abū Ḥāmid. *Faḍā'iḥ al-Bāṭiniyya*. Kuwait: Dār al-Kutub al-Thaqāfiyya, n.d.

al-Ghazālī, Abū Ḥāmid. *Fayṣal al-tafriqa bayna al-Islām wa-l-zandaqa*. Damascus: Muḥammad Bījū, 1993.

al-Ghazālī, Abū Ḥāmid. *The Incoherence of the Philosophers: A Parallel English-Arabic Text*. Edited and translated by Micheal E Marmura. Provo, UT: Brigham Young University Press, 2000.

al-Ghazālī, Abū Ḥāmid. *al-Iqtiṣād fī al-i'tiqād*. Cairo: Dār al-Baṣā'ir, 2009.

al-Ghazālī, Abū Ḥāmid. *al-Munqidh min al-ḍalāl*. Beirut: Dār al-Kutub al-'Ilmiyya, 1988.

al-Ghazālī, Abū Ḥāmid. *al-Mustaṣfā min 'ilm al-uṣūl*. Beirut: Dār al-Kutub al-'Ilmiyya, 1993.

al-Ghazālī, Abū Ḥāmid. *Qānūn al-ta'wīl*. Damascus: Muḥammad Bījū, 1992.

al-Ghazālī, Muḥammad. *Turāthunā al-fikrī fī mīzān al-shar' wa-l-'aql*. 8th ed. Cairo: Dār al-Shurūq, 2003.

al-Ghumārī, 'Abd Allāh. *Afḍal maqūl fī manāqib afḍal rasūl*. Cairo: Maktabat al-Qāhira, 2005.

Griffel, Frank. "Al-Ghazālī at His Most Rationalist: The Universal Rule for Allegorically Interpreting Revelation (*al-Qānūn al-Kullī fī t-Ta 'wīl*)." In *Islam and Rationality: The Impact of al-Ghazālī*, vol. 1, edited by Frank Griffel, 89–120. Leiden: Brill, 2015.

Grün, Rainer. "Direct Dating of Human Fossils." *American Journal of Physical Anthropology* 131, no. 43 (2006): 2–48.

Gurven, Michael, and Hillard Kaplan. "Longevity Among Hunter-Gatherers: A Cross-Cultural Examination." *Population and Development Review* 33, no. 2 (2007): 321–365.

al-Harawī, Abū Ismāʿīl. *al-Arbaʿīn fī dalāʾil al-tawḥīd*. Medina: n.p., 1984.

Ḥasan, Masʿūd. "Wafayāt: Dāktar Muḥammad al-Zubayr Ṣiddīqī." *Maʿārif* 177, no. 4 (1976): 382–397.

al-Ḥasanī, Mujīr al-Khaṭīb. *Maʿrifat madār al-isnād wa-bayān makānatihi fī ʿilm ʿilal al-ḥadīth*. Riyadh: Dār al-Maymān, 2007.

al-Haythamī, Nūr al-Dīn. *Majmaʿ al-zawāʾid wa-manbaʿ al-fawāʾid*. Cairo: Maktabat al-Qudsī, 1994.

al-Ḥāzimī, ʿIṣām. *Ṭūl abīnā Ādam ʿalayhi al-salām: shubuhāt wa-rudūd*. Riyadh: Waqf al-Itqān, 2020.

He, Qime, and Brian Morris. "Shorter Men Live Longer: Association of Height with Longevity and FOXO3 Genotype in American Men of Japanese Ancestry." *PLoS One* 9, no. 5 (2014): 1–8.

Hermanussen, Michael. "Stature of Early Europeans." *Hormones* 2, no. 3 (2003): 175–178.

Hershon, Paul. *A Talmudic Miscellany: A Thousand and One Extracts from the Talmud, the Midrashim and the Kabbalah*. Oxford: Routledge, 2000.

Hill, Carol. "Making Sense of the Numbers in Genesis." *Perspectives on Science and Christian Faith* 55, no. 4 (2003): 239–251.

Hublin, Jean-Jacques, Abdelouahed Ben-Ncer, Shara E. Bailey, Sarah E. Freidline, Simon Neubauer, Matthew M. Skinner, Inga Bergmann, et al. "New Fossils from Jebel Irhoud, Morocco and the Pan-African Origin of Homo Sapiens." *Nature* 546 (2017): 289–292.

al-Ḥuwaynī, Abū Isḥāq. *al-Manīḥa bi-silsilat al-aḥādīth al-ṣaḥīḥa*. Mansoura: Maktabat Ibn ʿAbbās, n.d.

al-Ḥuwaynī, Abū Isḥāq. *Tanbīh al-hājid ilā mā waqaʿa min al-naẓar fī kutub al-amājid*. Beirut: al-Maḥajja, n.d.

Ibn ʿAbd al-Barr, Yūsuf b. ʿAbd Allāh b. Muḥammad. *al-Tamhīd li-mā fī al-Muwaṭṭaʾ min al-maʿānī wa-l-asānīd*. Rabat: Wizārat ʿUmūm al-Awqāf wa-l-Shuʾūn al-Islāmiyya, 1412 AH.

Ibn Abī ʿĀṣim, Abū Bakr. *al-Sunna* . Beirut: al-Maktab al-Islāmī, 1400 AH.

Ibn Abī al-Dunyā, Abū Bakr. *Ṣifat al-janna*. Beirut: Muʾassasat al-Risāla, 1997.

Ibn ʿĀbidīn, Muḥammad Amīn. *Radd al-muḥtār ʿalā al-Durr al-mukhtār*. Damascus: Dār al-Thaqāfa al-Islāmiyya, 2011.

Ibn Abī Shayba, Abū Bakr. *Muṣannaf Ibn Abī Shayba*. Edited by Muḥammad ʿAwwāma. Jeddah: Dār al-Qibla, 2006.

Ibn Abī Shayba, ʿUthmān b. Muḥammad. *Suʾālāt ʿUthmān b. Muḥammad b. Abī Shayba li-l-Imām ʿAlī b. al-Madīnī*. Cairo: al-Fārūq al-Ḥadītha, n.d.

Ibn Abī Ṭālib, Makkī. *al-Hidāya ilā bulūgh al-nihāya*. Sharjah: Jāmi'at al-Shāriqa, 2008.

Ibn 'Adī, Abū Aḥmad. *al-Kāmil fī ḍu'afā' al-rijāl*. 7 vols. Beirut: Dār al-Fikr, 1985.

Ibn 'Adī, Abū Aḥmad. *al-Kāmil fī ḍu'afā' al-rijāl*. Beirut: Dār al-Kutub al-'Ilmiyya, 1997.

Ibn al-'Arabī, Abū Bakr. *Aḥkām al-Qur'ān*. Beirut: Dār al-Kutub al-'Ilmiyya, 2003.

Ibn 'Asākir, 'Alī b. al-Ḥasan. *Tārīkh madīnat Dimashq*. Beirut: Dār al-Fikr, 1995.

Ibn 'Āshūr, Muḥammad al-Ṭāhir. *Tafsīr al-Taḥrīr wa-l-tanwīr*. Tunis: al-Dār al-Tūnisiyya, 1984.

Ibn 'Aṭiyya, 'Abd al-Ḥaqq b. Ghālib. *al-Muḥarrar al-wajīz fī tafsīr al-Kitāb al-'Azīz*. Beirut: Dār al-Kutub al-'Ilmiyya, 1422 AH.

Ibn al-Bakhtarī, Abū Ja'far. *Majmū' fīhi muṣannafāt Abī Ja'far b. al-Bakhtarī*. Beirut: Dār al-Bashā'ir al-Islāmiyya, 2001.

Ibn Baṭṭāl, Abū al-Ḥasan 'Alī. *Sharḥ Ṣaḥīḥ al-Bukhārī*. Riyadh: Maktabat al-Rushd, 2003.

Ibn al-Ḍiyā', Abū al-Baqā'. *Tārīkh Makka al-musharrafa wa-l-masjid al-ḥarām wa-l-Madīna al-sharīfa wa-l-qabr al-sharīf*. Beirut: Dār al-Kutub al-'Ilmiyya, 2004.

Ibn Faḍlān, Aḥmad. *Ibn Faḍlān, Aḥmad Ibn Faḍlān and the Land of Darkness: Arab Travelers in the Far North*. Translated by Paul Lunde and Caroline Stone. London: Penguin Books, 2012.

Ibn Fūrak, Abū Bakr. *Mushkil al-ḥadīth wa-bayānuhu*. Beirut: 'Ālam al-Kutub, 1985.

Ibn Ḥadīda, Jamāl al-Dīn. *al-Miṣbāḥ al-muḍī fī kitāb al-nabī al-ummī wa-rusulihi ilā mulūk al-arḍ min 'arabī wa-'ajamī*. Beirut: 'Ālam al-Kutub, n.d.

Ibn Ḥajar al-'Asqalānī, Abū al-Faḍl. *Badhl al-mā'ūn fī faḍl al-ṭā'ūn*. Riyadh: Dār al-'Āṣima, n.d.

Ibn Ḥajar al-'Asqalānī, Abū al-Faḍl. *Fatḥ al-Bārī sharḥ Ṣaḥīḥ al-Bukhārī*. Edited by 'Abd al-'Azīz bin Bāz and Ayman Fu'ād 'Abd al-Bāqī. 14 vols. Beirut: Dār al-Kutub al-'Ilmiyya, 1997.

Ibn Ḥajar al-'Asqalānī, Abū al-Faḍl. *Lisān al-Mīzān*. Aleppo: Maktab al-Maṭbū'āt al-Islāmiyya, 2002.

Ibn Ḥajar al-'Asqalānī, Abū al-Faḍl. *Masā'il ajāba 'anhā al-Ḥāfiẓ Ibn Ḥajar al-'Asqalānī*. Cairo: Dār al-Imām Aḥmad, 2007.

Ibn Ḥajar al-'Asqalānī, Abū al-Faḍl. *al-Maṭālib al-'āliya bi-zawā'id al-Masānīd al-thamāniya*. Riyadh: Dār al-'Āṣima, 2000.

Ibn Ḥajar al-ʿAsqalānī, Abū al-Faḍl. *al-Qawl al-musaddad fī al-dhabb ʿan al-Musnad li-l-Imām Aḥmad*. Cairo: Maktabat Ibn Taymiyya, 1981.

Ibn Ḥajar al-ʿAsqalānī, Abū al-Faḍl. *Tahdhīb al-Tahdhīb*. Deccan: Maṭbaʿat Dāʾirat al-Maʿārif al-Niẓāmiyya, 1325 AH.

Ibn Ḥajar al-ʿAsqalānī, Abū al-Faḍl. *al-Zahr al-naḍir fī ḥāl al-Khaḍir*. New Delhi: Majmaʿ al-Buḥūth al-Islāmiyya, 1988.

Ibn Ḥajar al-Haytamī, Shihāb al-Dīn Aḥmad. *al-Fatāwā al-ḥadīthiyya*. Edited by Muḥammad ʿAbd al-Raḥmān al-Marʿashlī. Beirut: Dār Iḥyāʾ al-Turāth al-ʿArabī, 1998.

Ibn Ḥammād, Nuʿaym. *al-Fitan*. Cairo: Maktabat al-Tawḥīd, 1412 AH.

Ibn Ḥanbal, Aḥmad. *al-ʿIlal wa-maʿrifat al-rijāl*. Riyadh: Dār al-Khānī, 2001.

Ibn Ḥanbal, Aḥmad. *Musnad Aḥmad*. Edited by Shuʿayb al-Arnāʾūṭ et al. 50 vols. Beirut: Muʾassasat al-Risāla, 1995–2001.

Ibn Ḥazm, ʿAlī. *al-Iḥkām fī uṣūl al-aḥkām*. Beirut: Dār al-Āfāq al-Jadīda, 1983.

Ibn Hubayra, Yaḥyā. *al-Ifṣāḥ ʿan maʿānī al-Ṣiḥāḥ*. Riyadh: Dār al-Waṭan, 1996.

Ibn ʿIwaḍ Allāh, Ṭāriq. *Jāmiʿ al-masāʾil al-ḥadīthiyya*. Cairo: Dār Ibn ʿAffān, 2006.

Ibn al-Jawzī, Abū al-Faraj. *al-Ḍuʿafāʾ wa-l-matrūkūn*. Beirut: Dār al-Kutub al-ʿIlmiyya, 1986.

Ibn al-Jawzī, Abū al-Faraj. *Kitāb al-Quṣṣāṣ wa-l-mudhakkirīn*. Beirut: Dar El-Machreq, 1986.

Ibn al-Jawzī, Abū al-Faraj. *al-Mawḍūʿāt*. Medina: al-Maktaba al-Salafiyya, 1968.

Ibn al-Jawzī, Abū al-Faraj. *al-Muntaẓam fī tārīkh al-mulūk wa-l-umam*. Beirut: Dār al-Kutub al-ʿIlmiyya, 1992.

Ibn al-Jawzī, Abū al-Faraj. *Zād al-masīr*. Edited by ʿAbd al-Razzāq Mahdī. 4 vols. Beirut: Dār al-Kitāb al-ʿArabī, 1422/2002.

Ibn Kathīr, Ismāʿīl. *al-Bidāya wa-l-nihāya*. Beirut: Dār al-Fikr, 1986.

Ibn Kathīr, Ismāʿīl. *Tafsīr al-Qurʾān al-ʿAẓīm*. Riyadh: Dār Ṭayba, 1999.

Ibn Khaldūn, ʿAbd al-Raḥmān. *Dīwān al-mubtadaʾ wa-l-khabar*. Beirut: Dār al-Fikr, 1988.

Ibn Khaldūn, ʿAbd al-Raḥmān. *The Muqaddimah: An Introduction to History*. Edited and translated by Franz Rosenthal. Princeton: Princeton University Press, 1967.

Ibn al-Khaṭīb, Lisān al-Dīn. *Muqniʿat al-sāʾil ʿan al-maraḍ al-hāʾil*. Riyadh: Dār al-Amān, 2015.

Ibn Khuzayma, Muḥammad. *Ṣaḥīḥ Ibn Khuzayma*. Edited by Muḥammad Muṣṭafā al-Aʿẓamī. Beirut: al-Maktab al-Islāmī, 1992.

Ibn Khuzayma, Muḥammad. *al-Tawḥīd*. Riyadh: Dār al-Rushd, 1994.

Ibn Mandah, Muḥammad. *Min Amālī Ibn Mandah.* Damascus: al-Assad National Library, manuscript no. 3772.

Ibn Mandah, Muḥammad. *al-Radd ʿalā al-Jahmiyya.* Medina: Maktabat al-Ghurabāʾ al-Athariyya, 1994.

Ibn al-Mubārak, ʿAbd Allāh. *Kitāb al-Raqāʾiq.* Bahrain: Wizārat al-ʿAdl wa-l-Shuʾūn al-Islāmiyya, 2014.

Ibn al-Mulaqqin, ʿUmar b ʿAlī. *al-Tawḍīḥ li-sharḥ al-Jāmiʿ al-ṣaḥīḥ.* Damascus: Dār al-Nawādir, 2008.

Ibn Munabbih, Hammām. *Ṣaḥīfat Hammām b. Munabbih.* Beirut: al-Maktab al-Islāmī, 1987.

Ibn Qayyim al-Jawziyya, Muḥammad. *al-Manār al-munīf fī al-ṣaḥīḥ wa-l-ḍaʿīf.* Aleppo: Maktab al-Maṭbūʿāt al-Islāmiyya, 1970.

Ibn Qayyim al-Jawziyya, Muḥammad. *Miftāḥ dār al-saʿāda wa-manshūr wilāyat al-ʿilm wa-l-irāda.* Jeddah: Majmaʿ al-Fiqh al-Islāmī, 2010.

Ibn Qayyim al-Jawziyya, Muḥammad. *Tuḥfat al-mawdūd bi-aḥkām al-mawlūd.* Damascus: Dār al-Bayān, 1971.

Ibn Qayyim al-Jawziyya, Muḥammad. *Zād al-maʿād fī hady khayr al-ʿibād.* Beirut: Muʾassasat al-Risāla, 1994.

Ibn Qutayba, ʿAbd Allāh. *Taʾwīl mukhtalif al-ḥadīth.* Beirut: al-Maktab al-Islāmī, 1999.

Ibn Rajab al-Ḥanbalī, ʿAbd al-Raḥmān. *Ahwāl al-qubūr.* Edited by ʿĀṭif Ṣābir Shāhīn. Mansoura: Dār al-Ghad al-Jadīd, 2005.

Ibn Rajab al-Ḥanbalī, ʿAbd al-Raḥmān. *Fatḥ al-Bārī fī sharḥ Ṣaḥīḥ al-Bukhārī.* Cairo: Maktabat Taḥqīq Dār al-Ḥaramayn, 1996.

Ibn Rajab al-Ḥanbalī, ʿAbd al-Raḥmān. *Sharḥ ʿIlal al-Tirmidhī.* Zarqa: Maktabat al-Manār, 1987.

Ibn al-Ṣalāḥ, Abū ʿAmr ʿUthmān. *Maʿrifat anwāʿ ʿilm al-ḥadīth.* Beirut: Dār al-Fikr, 1986.

Ibn al-Ṣalāḥ, Abū ʿAmr ʿUthmān. "Risāla fī waṣl al-balāghāt al-arbaʿa fī *al-Muwaṭṭa*ʾ." In *Khams rasāʾil fī ʿulūm al-ḥadīth,* edited by ʿAbd al-Fattāḥ Abū Ghudda, 179–212. Beirut: Dār al-Bashāʾir al-Islāmiyya, 2010.

Ibn Sallām, Yaḥyā. *Tafsīr Yaḥyā b. Sallām.* Beirut: Dār al-Kutub al-ʿIlmiyya, 2004.

Ibn Taymiyya, Taqī al-Dīn. *Darʾ taʿāruḍ al-ʿaql wa-l-naql.* Riyadh: Jāmiʿat al-Imām, 1991.

Ibn Taymiyya, Taqī al-Dīn. *Majmūʿ fatāwā Shaykh al-Islām Aḥmad b. Taymiyya.* Medina: Majmaʿ al-Malik Fahd, 1995.

al-Idlibī, Ṣalāḥ al-Dīn. *Manhaj naqd al-matn ʿinda ʿulamāʾ al-ḥadīth al-nabawī.* Amman: Dār al-Fatḥ, 2013.

ʿĪdū, ʿIṣām. *Nashʾat ʿilm al-muṣṭalaḥ wa-l-ḥadd al-fāṣil bayna al-mutaqaddimīn wa-l-mutaʾakhkhirīn.* Amman: Arwiqa, 2016.

al-ʿIrāqī, Abū Zurʿa. *Ṭarh al-tathrīb fī sharḥ al-Taqrīb.* Beirut: Dār Iḥyāʾ al-Turāth al-ʿArabī, n.d.

al-Ithyūbī, Muḥammad. *Sharḥ Sunan al-Nasāʾī.* Riyadh: Dār al-Miʿrāj, 1996.

Jackson, Sherman. *On the Boundaries of Theological Tolerance in Islam: Abū Ḥāmid al-Ghazālī's* Fayṣal al-Tafriqa. Oxford: Oxford University Press, 2002.

Jamīl, Abū Sāra Farīd. *Athar al-ʿilm al-tajrībī fī kashf naqd al-ḥadīth al-nabawī.* Beirut: Markaz Namāʾ, 2016.

Jawnpūrī, Yūnus. *al-Nibrās al-sārī ilā riyāḍ al-Bukhārī.* Gujrat: Maktabat al-Qalam, n.d.

al-Jazāʾirī, Ṭāhir. *Tawjīh al-naẓar ilā uṣūl al-athar.* Aleppo: Maktab al-Maṭbūʿāt al-Islāmiyya, 1995.

Juynboll, Gautier H.A. *The Authenticity of the Tradition Literature: Discussions in Modern Egypt.* Leiden: Brill, 1969.

al-Jūzajānī, Abū Isḥāq. *Aḥwāl al-rijāl.* Faisalabad: Hadith Academy, n.d.

al-Kalābādhī, Abū Bakr. *Baḥr al-fawāʾid.* Cairo: Dār al-Salām, 2008.

Kalin, Ibrahim. *Reason and Rationality in the Qurʾān.* Amman: Kalam Research & Media, 2015.

Kalin, Ibrahim. *Islam and Science.* n.d. http://www.oxfordislamicstudies.com/Public/focus/essay1009_science.html.

al-Kandarī, Jihād, and Ashraf al-Quḍāh. "Ḥadīth al-ṣūra: khalaqa Allāh Ādam ʿalā ṣūratihī: dirāsa naqdiyya." *IUG Journal of Islamic Studies* 27, no. 1 (2019): 432–445.

Kāndhlawī, Muḥammad Zakariyyā. *al-Abwāb wa-l-tarājim.* Beirut: Dār al-Bashāʾir al-Islāmiyya, 2012.

Kashmīrī, Anwar Shāh. *al-ʿArf al-shadhī sharḥ Sunan al-Tirmidhī.* Beirut: Dār al-Turāth al-ʿArabī, 2004.

Kashmīrī, Anwar Shāh. *Fayḍ al-Bārī bi-sharḥ Ṣaḥīḥ al-Bukhārī.* Beirut: Dār al-Kutub al-ʿIlmiyya, 2005.

al-Kattānī, ʿAbd al-Ḥayy. *Fihris al-fahāris wa-l-athbāt wa-muʿjam al-maʿājim wa-l-mashyakhāt wa-l-musalsalāt.* Beirut: Dār al-Gharb al-Islāmī, 1982.

al-Kattānī, Muḥammad b. Jaʿfar. *Naẓm al-mutanāthir min al-ḥadīth al-mutawātir.* Cairo: Dār al-Kutub al-Salafiyya, n.d.

al-Kawtharī, Muḥammad Zāhid. *Ḥusn al-taqāḍī fī sīrat al-Imām Abī Yūsuf al-qāḍī*. Amman: Dār al-Fatḥ, 2017.

al-Kawtharī, Muḥammad Zāhid. *Maqālāt al-Kawtharī*. Cairo: al-Maktaba al-Azhariyya, 1994.

Kazābar, ʿIzz al-Dīn. *Ṭūl Ādam wa-l-insān wa-munḥanā nuqṣānihi maʿa al-zamān wa-l-radd ʿalā ʿAdnān*. December 19, 2012. http://kazaaber.blogspot.com/2012/12/blog-post.html.

al-Khanbarjī, Muḥammad. "Ikhtiṣār al-matn wa-manhaj al-Imām al-Bukhārī fīhi." PhD dissertation, University of Jordan, 2010.

al-Khaṭīb al-Baghdādī, Abū Bakr. *al-Jāmiʿ li-akhlāq al-rāwī wa-ādāb al-sāmiʿ*. Riyadh: Maktabat al-Maʿārif, 1983.

al-Khatlī, Isḥāq b. Ibrāhīm. *Kitāb al-Dībāj*. Beirut: Dār al-Bashāʾir, 1994.

al-Khaṭṭābī, Abū Sulaymān. *Maʿālim al-sunan*. 2nd ed. 4 vols. Beirut: al-Maktaba al-ʿIlmiyya, 1981.

al-Khaṭṭābī, Ḥamd b. Muḥammad. *Aʿlām al-ḥadīth fī sharḥ maʿānī Kitāb al-Jāmiʿ al-ṣaḥīḥ*. Mecca: Jāmiʿat Umm al-Qurā, 1988.

al-Khaṭṭābī, Ḥamd b. Muḥammad. *Aʿlām al-sunan*. Mecca: Jāmiʿat Umm al-Qurā, 1988.

al-Khinn, Muḥammad. *al-Qaṭʿī wa-l-ẓannī fī al-thubūt wa-l-dalāla ʿinda al-uṣūliyyīn*. Damascus: Dār al-Kalim al-Ṭayyib, 2007.

al-Kinānī, Ibn ʿIrāq. *Tanzīh al-sharīʿa al-marfūʿa ʿan al-akhbār al-shanīʿa al-mawḍūʿa*. Beirut: Dār al-Kutub al-ʿIlmiyya, 1399 AH.

al-Kirmānī, Muḥammad b. Yūsuf. *al-Kawākib al-darārī fī sharḥ Saḥīḥ al-Bukhārī*. Beirut: Dār Iḥyāʾ al-Turāth al-ʿArabī, 1937.

Kister, Meir Jacob. "'Ḥaddithū ʿan banī isrāʾīl wa-lā ḥaraj': A Study of an Early Tradition." *Israel Oriental Studies* 2 (1979): 215–239.

Kugel, James L. *Traditions of the Bible: A Guide to the Bible as It Was at the Start of the Common Era*. Cambridge, MA: Harvard University Press, 1998.

Ladyman, James. *Understanding Philosophy of Science*. London: Routledge, 2002.

Laher, Suheil. "Twisted Threads: Genesis, Development and Application of the Term and Concept of Tawatur in Islamic Thought." PhD dissertation, Harvard University, 2014.

al-Laknawī, ʿAbd al-Ḥayy. *Ẓafar al-amānī bi-sharḥ Mukhtaṣar al-Sayyid al-Sharīf al-Jurjānī*. Aleppo: Maktab al-Maṭbūʿāt al-Islāmiyya, 1416 AH.

Lāshīn, Mūsā. *Fatḥ al-Munʿim sharḥ Ṣaḥīḥ Muslim*. Cairo: Dār al-Shurūq, 2002.

Lewy, Hildegard. "Origin and Development of the Sexagesimal System of Numeration." *Journal of the American Oriental Society* 69, no. 1 (1949): 1–11.

al-Maḥbūbī, ʿUbayd Allāh. *al-Tawḍīḥ sharḥ al-Tanqīḥ*. Cairo: Maktabat Ṣabīḥ, n.d.

al-Makkī, Abū Ṭālib. *Qūt al-qulūb*. 2 vols. in 1. Cairo: Maṭbaʿat al-Anwār al-Muḥammadiyya, 1985.

Malik, Shoaib Ahmed. *Atheism and Islam: A Contemporary Discourse*. Abu Dhabi: Kalam Research & Media , 2018.

Mālik b. Anas. *al-Muwaṭṭa'*. Abu Dhabi: Mu'assasat Zāyid b. Sulṭān, 2004.

al-Manṣūrī, Nāyif b. Ṣalāḥ. *Irshād al-qāṣī wa-l-dānī ilā tarājim shuyūkh al-Ṭabarānī*. Sharjah: Maktabat Ibn Taymiyya, 2006.

al-Maqdisī, al-Ḍiyā'. *al-Muntaqā min masmūʿāt Marw*. Damascus: al-Assad National Library, manuscript no. 344.

al-Maqdisī, Muḥammad b. Ṭāhir. *Aṭrāf al-gharā'ib wa-l-afrād*. Beirut: Dār al-Kutub al-ʿIlmiyya, 1997.

al-Maqdisī, Muḥammad b. Ṭāhir. *Dhakhīrat al-ḥuffāẓ al-mukharraj ʿalā al-ḥurūf wa-l-alfāẓ*. Riyadh: Dār al-Salaf, 1996.

al-Maqdisī, Muṭahhar b. Ṭāhir. *Kitāb al-Bad' wa-l-tārīkh*. Cairo: Maktabat al-Thaqāfa al-Dīniyya, n.d.

al-Maqdisī, Muṭahhar b. Ṭāhir. *Kitāb al-Bad' wa-l-tārīkh*. Istanbul: Süleymaniye Yazma Eser Library, Dāmād Ibrāhīm Pāshā, manuscript no. 918.

al-Maqrīzī, Aḥmad. *al-Mawāʿiẓ wa-l-iʿtibār bi-dhikr al-khiṭaṭ wa-l-āthār*. Beirut: Dār al-Kutub al-ʿIlmiyya, 1997.

Marck, Adrien, Juliana Antero, Geoffroy Berthelot, Guillaume Saulière, Jean-Marc Jancovici, Valérie Masson-Delmotte, Gilles Bœuf, Michael Spedding, Eric Le Bourg, and Jean-François Toussaint. "Are We Reaching the Limits of Homo Sapiens?" *Frontiers in Physiology* 8 (2017): 1–12.

al-Marwazī, Isḥāq b. Manṣūr al-Kawsaj. *Masā'il al-Imām Aḥmad b. Ḥanbal wa-Isḥāq b. Rāhawayh*. Medina: al-Jāmiʿa al-Islāmiyya, 2002.

al-Marwazī, Muḥammad b. Naṣr. *Taʿẓīm qadr al-ṣalāh*. Medina: Maktabat al-Dār, 1986.

Mathers, Kathryn, and Maciej Henneberg. "Were We Ever That Big? Gradual Increase in Hominid Body Size Over Time." *Homo* 46, no. 2 (1995): 141–173.

Matt, Daniel C., trans. *The Zohar: Pritzker Edition*. Redwood City, CA: Stanford University Press, 2005.

al-Māturīdī, Abū Manṣūr. *Ta'wīlāt ahl al-sunna*. Beirut: Dār al-Kutub al-ʿIlmiyya, 2005.

Mayor, Adrienne. *The First Fossil Hunters: Dinosaurs, Mammoths, and Myth in Greek and Roman Times*. Princeton: Princeton University Press, 2011.

Mayor, Adrienne. *Fossil Legends of the First Americans*. Princeton: Princeton University Press, 2005.

al-Māzarī, Muḥammad. *al-Muʿlim bi-fawāʾid Muslim*. Tunis: al-Dār al-Tūnisiyya, 1991.

Melchert, Christopher. "God Created Adam in His Image." *Journal of Qurʾanic Studies* 13, no. 1 (2011): 113–124.

al-Mizzī, Jamāl al-Dīn. *Tahdhīb al-kamāl fī asmāʾ al-rijāl*. Edited by Bashshār ʿAwwād Maʿrūf. Beirut: Muʾassasat al-Risāla, 1983.

al-Mizzī, Jamāl al-Dīn. *Tuḥfat al-ashrāf bi-maʿrifat al-aṭrāf*. Edited by Bashshār ʿAwwād Maʿrūf. Beirut: Dār al-Gharb al-Islāmī, 1999.

Montagu, J. "Length of Life in the Ancient World: A Controlled Study." *Journal of the Royal Society of Medicine* 87 (1994): 25–26.

Moog, Florence. "Gulliver Was a Bad Scientist." *Scientific American* 179, no. 5 (1948): 52–55.

al-Muʿallimī, ʿAbd al-Raḥmān. *al-Anwār al-kāshifa li-mā fī kitāb Aḍwāʾ ʿalā al-sunna min al-zalal wa-l-taḍlīl wa-l-mujāzafa*. Jeddah: Majmaʿ al-Fiqh al-Islāmī, 2012.

al-Muʿallimī, ʿAbd al-Raḥmān. *Ḥawla tafsīr al-Fakhr al-Rāzī wa-takmilatihi*. Mecca: Dār ʿĀlam al-Fawāʾid, 1424 AH.

al-Muʿallimī, ʿAbd al-Raḥmān. *al-Qāʾid ilā taṣḥīḥ al-ʿaqāʾid*. Beirut: al-Maktab al-Islāmī, 1984.

Mubārakpūrī, ʿAbd al-Raḥmān. *Tuḥfat al-aḥwadhī bi-sharḥ Jāmiʿ al-Tirmidhī*. Beirut: Dār al-Fikr, n.d.

Mughal, Muḥammad Zāhid. "ʿAql wa-naql kā taʿaruḍ awr taʾwīl key liey Imām Ghazālī kā qānūn kullī." *al-Sharīʿa* 31, no. 5 (2020): 46–52.

al-Munāwī, ʿAbd al-Raʾūf. *Fayḍ al-Qadīr sharḥ al-Jāmiʿ al-ṣaghīr*. Cairo: al-Maktaba al-Tijāriyya al-Kubrā, n.d.

al-Munāwī, ʿAbd al-Raʾūf. *al-Taysīr bi-sharḥ al-Jāmiʿ al-ṣaghīr*. Riyadh: Maktabat al-Imām al-Shāfiʿī, 1988.

Muslim b. al-Ḥajjāj. *Kitāb al-Tamyīz*. Riyadh: Sharikat al-Ṭibāʿa al-ʿArabiyya, 1982.

Muslim b. al-Ḥajjāj. *al-Kunā wa-l-asmāʾ*. Medina: al-Jāmiʿa al-Islāmiyya, 1984.

Muslim b. al-Ḥajjāj. *al-Musnad al-ṣaḥīḥ al-mukhtaṣar min al-sunan bi-naql al-ʿadl ʿan al-ʿadl ʿan Rasūl Allāh*. Jeddah: Dār al-Minhāj, 2013.

al-Muṭayrī, Mastūra Rajā. "al-Tanāẓur wa-l-tabāyun bayna ikhtiṣār al-ḥadīth wa-taqṭīʿihi ʿinda al-muḥaddithīn." *Majallat al-Dirāsa al-Islāmiyya* 89, no. 13 (2018): 414–438.

al-Muṭīʿī, Muḥammad Bakhīt. *Tawfīq al-Raḥmān li-l-tawfīq bayna mā qālahu ʿulamāʾ al-hayʾa wa-bayna mā jāʾa fī al-aḥādīth al-ṣaḥīḥa wa-āyāt al-Qurʾān*. Beirut: Dār al-Minhāj, 2016.

al-Nadwī, Muḥammad Akram. *al-Farāʾid fī ʿawālī al-asānīd wa-ghawālī al-fawāʾid*. Beirut: Dār al-Bashāʾir al-Islāmiyya, 2015.

Naeem, Fuad. "The Imaginal World (*ʿĀlam al-Mithāl*) in the Philosophy of Shāh Walī Allāh Dihlawī." *Islamic Studies* 4, no. 3 (2006): 363–390.

al-Nahhām, Ṣāliḥ Sālim. *al-Ikhtilāf al-uṣūlī fī al-tarjīḥ bi-kathrat al-adilla wa-l-ruwāh wa-atharuhu*. Doha: Wizārat al-Awqāf wa-l-Shuʾūn al-Islāmiyya, 2011.

al-Nājī, Burhān al-Dīn. *Ḥuṣūl al-bughya li-l-sāʾil hal li-aḥad fī al-janna liḥya*. Beirut: Dār al-Bashāʾir al-Islāmiyya, 2004.

al-Nasafī, Ḥāfiẓ al-Dīn. *Kashf al-asrār sharḥ al-muṣannif ʿalā al-Manār*. Beirut: Dār al-Kutub al-ʿIlmiyya, n.d.

al-Nasāʾī, Aḥmad b. Shuʿayb. *al-Sunan al-kubrā*. Edited by Shuʿayb al-Arnāʾūṭ. Beirut: Muʾassasat al-Risāla, 2002.

al-Nawawī, Yaḥyā b. Sharaf. *al-Minhāj fī sharḥ Ṣaḥīḥ Muslim b. al-Ḥajjāj*. Beirut: Dār Iḥyāʾ al-Turāth al-ʿArabī, 1972.

al-Nawawī, Yaḥyā b. Sharaf. *Tahdhīb al-asmāʾ wa-l-lughāt*. Beirut: Dār al-Kutub al-ʿIlmiyya, n.d.

al-Nuʿmānī, ʿAbd al-Rashīd. *Imām Ibn Mājah awr ʿilm-e ḥadīth*. Karachi: Aṣaḥḥ al-Maṭābiʿ, n.d.

al-Nuʿmānī, ʿAbd al-Rashīd. *al-Imām Ibn Mājah wa-kitābuhu al-Sunan*. Beirut: Dār al-Bashāʾir al-Islāmiyya, 1998.

Page, Don. "The Height of a Giraffe." *Foundations of Physics* 39, no. 10 (2009): 1097–1108.

Pavlovitch, Pavel. *The Formation of the Islamic Understanding of Kalâla in the Second Century AH (718–816 CE): Between Scripture and Canon*. Leiden: Brill, 2016.

Press, William H. "Man's Size in Terms of Fundamental Constants." *American Journal of Physics* 48, no. 8 (1980): 597.

al-Qāḍī ʿIyāḍ. *al-Ilmāʿ ilā maʿrifat uṣūl al-riwāya wa-taqyīd al-samāʿ*. Cairo: Maktabat Dār al-Turāth, 1970.

al-Qāḍī ʿIyāḍ. *Kitāb al-Shifāʾ*. Manama: Maktabat Niẓām Yaʿqūbī, 2015.

al-Qārī, Mullā ʿAlī. *al-Maṣnūʿ fī maʿrifat al-ḥadīth al-mawḍūʿ*. Beirut: Dār al-Bashāʾir al-Islāmiyya, 1994.

al-Qārī, Mullā ʿAlī. *Mirqāt al-mafātīḥ sharḥ Mishkāt al-Maṣābīḥ*. Beirut: Dār al-Fikr, 2002.

Quadri, Junaid. *Transformations of Tradition: Islamic Law in Colonial Modernity.* Oxford: Oxford University Press, 2021.

al-Quḍāh, Sharaf Maḥmūd. "al-Islām wa-l-ʿilm fī al-Qurʾān wa-l-sunna." *Majallat Kulliyyat al-Sharīʿa* 14 (1996).

Qureshi, Omar, and Aasim Padela. "When Must a Patient Seek Healthcare? Bridging the Perspectives of Islamic Jurists and Clinicians into Dialogue." *Zygon* 51, no. 3 (2016): 605–617.

al-Qurṭubī, Abū ʿAbd Allāh. *al-Jāmiʿ li-aḥkām al-Qurʾān.* Cairo: Dār al-Kutub al-Miṣriyya, 1964.

al-Qurṭubī, Abū ʿAbd Allāh. *al-Jāmiʿ li-aḥkām al-Qurʾān.* Edited by Muḥammad Ibrāhīm al-Ḥifnāwī and Maḥmūd Ḥāmid ʿUthmān. 20 vols. Cairo: Dār al-Ḥadīth, 1994.

al-Qurṭubī, Abū al-ʿAbbās. *al-Mufhim li-mā ashkala min talkhiṣ Kitāb Muslim.* Beirut: Dār Ibn Kathīr, 1996.

al-Rāzī, Abū Zurʿa. *Kitāb al-Ḍuʿafāʾ.* Edited by Saʿdī al-Hāshimī. Medina: al-Majlis al-ʿIlmī, 1982.

al-Razī, Fakhr al-Dīn. *Mafātīḥ al-ghayb.* Beirut: Dār Iḥyāʾ al-Turāth al-ʿArabī, 1999.

al-Razī, Fakhr al-Dīn. *The Great Exegesis (al-Tafsīr al-kabīr),* vol. 1, *The Fātiḥa.* Translated by Sohaib Saeed. Cambridge: The Royal Aal al-Bayt Institute for Islamic Thought and The Islamic Texts Society, 2018.

al-Rāzī, Ibn Abī Ḥātim. *ʿIlal al-ḥadīth.* Riyadh: Maṭbaʿat al-Ḥumayḍī, 2006.

al-Rāzī, Ibn Abī Ḥātim. *Kitāb al-Jarḥ wa-l-taʿdīl.* Deccan: Dāʾirat al-Maʿārif al-ʿUthmāniyya, 1952.

Riḍā, Rashīd. "Masʾalat inshiqāq al-qamar." *Majallat al-manār* 30 (1348 AH): 261–272.

Riḍā, Rashīd. *Tafsīr al-Manār.* Cairo: Dār al-Manār, 1948.

Romano, Marco, and Marco Avanzini. "The Skeletons of Cyclops and Lestrigons: Misinterpretation of Quaternary Vertebrates as Remains of the Mythological Giants." *Historical Biology* 31, no. 2 (2019): 117–139.

Rose, Mark. "When Giants Roamed the Earth." *Archaeological Institute of America* 58, no. 6 (2005).

Rosenstock, Eva, Julia Ebert, Robert Martin, Andreas Hicketier, Paul Walter, and Marcus Groß. "Human Stature in the Near East and Europe ca. 10,000–1000 BC: Its Spatiotemporal Development in a Bayesian Errors-in-Variables Model." *Archaeological and Anthropological Sciences* 11 (2019): 5657–5690.

Ruff, Christopher B., Erik Trinkaus, and Trenton W. Holliday. "Body Mass and Encephalization in Pleistocene Homo." *Nature* 387 (1997): 173–176.

al-Rūmī, Fahd. *Manhaj al-madrasa al-ʿaqliyya al-ḥadītha fī al-tafsīr*. Beirut: Muʾassasat al-Risāla, 1983.

Saḥnūn, Ibn Saʿīd al-Tanūkhī. *al-Mudawwana al-kubrā*. Beirut: Dār al-Kutub al-ʿIlmiyya, 1994.

al-Sakhāwī, Shams al-Dīn. *al-Maqāṣid al-ḥasana fī al-aḥādīth al-mushtahira ʿalā al-alsina*. Beirut: Dār al-Kitāb al-ʿArabī, 1985.

al-Samʿānī, Abū al-Muẓaffar. *Qawāṭiʿ al-adilla fī al-uṣūl*. Beirut: Dār al-Kutub al-ʿIlmiyya, 1999.

al-Sāmarrāʾī, Fāḍil Ṣāliḥ. *Maʿānī al-abniya al-ʿarabiyya*. Amman: Dār ʿAmmār, 2007.

al-Sāmarrāʾī, Fāḍil Ṣāliḥ. *Maʿānī al-naḥw*. Amman: Dār al-Fikr, 2000.

Samaras, Thomas. "How Height Is Related to Our Health and Longevity: A Review." *Nutrition and Health* 21, no. 4 (2012): 247–261.

Samaras, Thomas. "Human Scaling and the Body Mass Index." In *Human Body Size and the Laws of Scaling, Physiological, Performance, Growth, Longevity and Ecological Ramifications*, edited by Thomas Samaras, 1–15. New York: Nova Science Publishers, 2007.

Samaras, Thomas, Harold Elrick, and Lowell Storms. "Is Height Related to Longevity?" *Life Science* 72, no. 16 (2003): 1781–1802.

al-Samarqandī, Muḥyī al-Dīn. *Naqd matn al-ḥadīth fī ḍawʾ natāʾij al-ʿulūm al-tajrībiyya*. Beirut: Dār al-Kutub al-ʿIlmiyya, 2008.

al-Ṣanʿānī, ʿAbd al-Razzāq b. Hammām. *Muṣannaf ʿAbd al-Razzāq*. Beirut: al-Majlis al-ʿIlmī, 1982.

al-Ṣanʿānī, ʿAbd al-Razzāq b. Hammām. *Tafsīr ʿAbd al-Razzāq*. Riyadh: Maktabat al-Rushd, 1989.

al-Ṣanʿānī, al-Amīr Muḥammad. *al-Tanwīr sharḥ al-Jāmiʿ al-ṣaghīr*. Riyadh: Maktabat Dār al-Salām, 2011.

al-Saʿūd, Sulaymān b. ʿAbd Allāh. "Ikhtiṣār al-ḥadīth wa-atharuhu fī al-ruwāh wa-l-marwiyyāt: dirāsa waṣfiyya taḥlīliyya." *Majallat al-Jāmiʿa al-Islāmiyya* 9, no. 83 (1439 AH): 170–276.

Scott, Robert. "The Hebrew Cubit." *Journal of Biblical Literature* 77, no. 3 (1958): 205–214.

Shākir, Aḥmad. *al-Bāʿith al-ḥathīth sharḥ Ikhtiṣār ʿulūm al-ḥadīth*. Riyadh: Maktabat al-Maʿārif, 1996.

Shaltūt, Maḥmūd. *Fatāwā*. Cairo: Dār al-Shurūq, 2001.

al-Shaybānī, Muḥammad b. al-Ḥasan. *al-Aṣl*. Doha: Wizārat al-Awqāf, 2012.

Shea, John. *Prehistoric Stone Tools of Eastern Africa: A Guide.* New York: Cambridge University Press, 2020.

al-Shūlī, Wasīm. "Manhaj al-Bukhārī fī al-ḥadīth al-mudraj." *al-Majalla al-Dawliyya* 6, no. 7 (2016): 1–35.

Siddiqi, Muhammad Zubayr. *Hadith Literature: Its Origin, Development & Special Features.* Cambridge: Islamic Texts Society, 1993.

al-Silafī, Abū Ṭāhir. *al-Ṭuyūriyyāt.* Riyadh: Maktabat Aḍwā' al-Salaf, 2004.

al-Ṣimādī, ʿImād. "Maqāṣid al-Bukhārī fī riwāyat al-aḥādīth fī ghayr maẓānnihā." PhD dissertation, Jāmiʿat al-ʿUlūm al-Islāmiyya al-ʿĀlamiyya (Amman), 2017.

Sindhī, ʿUbayd Allāh. *Risāla fī muṣṭalaḥ al-ḥadīth.* Karachi: Ghulām Muṣṭafā, n.d.

Sindhī, ʿUbayd Allāh. *Sharḥ Ḥujjat Allāh.* Lahore: Maktabat Bayt al-Ḥikma, 1950.

al-Sindī, Abū al-Ḥasan. *Ḥāshiyat Ṣaḥīḥ al-Bukhārī.* Lahore: Maktaba Raḥmāniyya, n.d.

Stenmark, Mikael. *Scientism: Science, Ethics and Religion.* New York: Routledge, 2018.

Stulp, Gert, and Louise Barrett. "Evolutionary Perspectives on Human Height Variation." *Biological Reviews Cambridge Philosophical Society* 91, no. 1 (2016): 206–235.

Styne, Dennis, and Henry McHenry. "The Evolution of Stature in Humans." *Hormone Research* 39 (1993): 3–6.

al-Subkī, Tāj al-Dīn. *Ṭabaqāt al-Shāfiʿiyya al-kubrā.* Cairo: ʿĪsā al-Bābī al-Ḥalabī, 1964.

al-Suyūṭī, Jalāl al-Dīn. *al-Durr al-manthūr fī al-tafsīr bi-l-ma'thūr.* Beirut: Dār al-Fikr, 2011.

al-Suyūṭī, Jalāl al-Dīn. *Muʿtarak al-aqrān fī iʿjāz al-Qur'ān.* Beirut: Dār al-Kutub al-ʿIlmiyya, 1988.

al-Suyūṭī, Jalāl al-Dīn. *Rafʿ sha'n al-ḥubshān.* Edited by Muḥammad ʿAbd al-Wahhāb Faḍl. Cairo: self-published, 1991.

al-Suyūṭī, Jalāl al-Dīn. *Tadrīb al-rāwī fī sharḥ Taqrīb al-Nawāwī.* Edited by Muḥammad ʿAwwāma. Jeddah: Dār al-Minhāj, 2016.

al-Ṭabarānī, Abū al-Qāsim. *al-Muʿjam al-awsaṭ.* Cairo: Dār al-Ḥaramayn, 1415 AH.

al-Ṭabarānī, Abū al-Qāsim. *al-Muʿjam al-kabīr.* Cairo: Maktabat Ibn Taymiyya, n.d.

al-Ṭabarānī, Abū al-Qāsim. *al-Muʿjam al-ṣaghīr.* Beirut: al-Maktab al-Islāmī, 1985.

al-Ṭabarānī, Abū al-Qāsim. *Musnad al-Shāmiyyīn*. Beirut: Mu'assasat al-Risāla, 1986.

al-Ṭabarī, Ibn Jarīr. *Jāmi' al-bayān fī ta'wīl āy al-Qur'ān*. Beirut: Mu'assasat al-Risāla, 2000.

al-Ṭabarī, Ibn Jarīr. *Tārīkh al-rusul wa-l-mulūk*. Beirut: Dār al-Turāth, 1967.

al-Ṭabbākh, Muḥammad Rāghib. *Dhū al-Qarnayn wa-sadd al-Ṣīn: man huwa wa-ayna huwa?* Kuwait: Ghirās, 2003.

al-Taftāzānī, Sa'd al-Dīn, et al. *Majmū'at al-ḥawāshī al-bahiyya 'alā sharḥ al-'Aqā'id al-Nasafiyya*. 4 vols. Cairo: Maṭba'at Kurdistān, 1329 AH.

al-Ṭaḥāwī, Abū Ja'far. *Sharḥ ma'ānī al-āthār*. Beirut: 'Ālam al-Kutub, 1994.

al-Ṭaḥāwī, Abū Ja'far. *Sharḥ mushkil al-āthār*. Edited by Shu'ayb al-Arnā'ūṭ. 16 vols. Beirut: Mu'assasat al-Risāla, 1994.

Tammām, Abū al-Qāsim. *Fawā'id*. Edited by Ḥamdī 'Abd al-Majīd al-Salafī. 2 vols. Riyadh: Maktabat al-Rushd, 1412/2002.

al-Ṭayyār, Musā'id. *al-Taḥrīr fī uṣūl al-tafsīr*. Jeddah: Ma'had al-Imām al-Shāṭibī, 2014.

al-Tha'labī, Aḥmad b. Muḥammad. *al-Kashf wa-l-bayān*. Jeddah: Dār al-Tafsīr, 2015.

Thureau-Dangin, François. "Sketch of a History of the Sexagesimal System." *Osiris* 7 (1939): 95–141.

al-Ṭībī, Sharaf al-Dīn. *al-Kāshif 'an ḥaqā'iq al-sunan*. Riyadh: Maktabat al-Bāz, 1997.

al-Tirmidhī, Abū 'Īsā Muḥammad. *al-Jāmi' al-kabīr*. Edited by Bashshār 'Awwād Ma'rūf. Beirut: Dār al-Gharb al-Islāmī, 1998.

al-Tirmidhī, al-Ḥakīm Abū 'Abd Allāh. *Nawādir al-uṣūl fī aḥādīth al-Rasūl*. Beirut: Dār al-Jīl, 1992.

Tottoli, Roberto. "Origin and Use of the Term *Isrā'īliyyāt* in Muslim Literature." *Arabica* 46, no. 2 (1999): 194–201.

al-Ṭūfī, Najm al-Dīn Sulaymān. *al-Iksīr fī 'ilm al-tafsīr*. Beirut: Dār al-Awzā'ī, 1989.

al-Ṭūfī, Najm al-Dīn Sulaymān. *Sharḥ Mukhtaṣar al-Rawḍa*. Beirut: Mu'assasat al-Risāla, 1987.

al-Turkumānī, 'Abd al-Majīd. *Dirāsāt fī uṣūl al-ḥadīth 'alā manhaj al-Ḥanafiyya*. Karachi: Madrasat al-Nu'mān, 2009.

al-Tuwayjirī, Ḥamūd b. 'Abd Allāh. *'Aqīdat ahl al-īmān fī khalq Ādam 'alā ṣūrat al-Raḥmān*. Riyadh: Dār al-Liwā', 1989.

al-ʿUmarī, Muḥammad. *al-Nubuwwa bayna al-mutakallimīn wa-l-falāsifa wa-l-ṣūfiyya*. Amman: Dār al-Fatḥ, 2015.

al-ʿUqaylī, Abū Jaʿfar. *Kitāb al-Ḍuʿafāʾ al-kabīr*. Beirut: Dār al-Maktaba al-ʿIlmiyya, 1984.

al-ʿUtaybī, Ghāzī. "al-Tarjīḥ bi-kathrat al-ruwāh: dirāsa uṣūliyya taṭbīqiyya." *Majallat Jāmiʿat Umm al-Qurā li-ʿUlūm al-Sharīʿa wa-l-Dirāsāt al-Islāmiyya* 44 (1429 AH): 297–364.

ʿUthmānī, Muḥammad Shafīʿ. *Awzān-e sharʿiyya*. Karachi: Idārat al-Maʿārif, n.d.

ʿUthmānī, Muḥammad Taqī. *Takmilat Fatḥ al-Mulhim*. Beirut: Dār Iḥyāʾ al-Turāth al-ʿArabī, 2006.

ʿUthmānī, Muḥammad Taqī. *ʿUlūm al-Qurʾān*. Karachi: Maktabat Dār al-ʿUlūm, n.d.

ʿUthmānī, Shabbīr Aḥmad. *Fatḥ al-Mulhim bi-sharḥ Ṣaḥīḥ al-Imām Muslim*. Beirut: Dār Iḥyāʾ al-Turāth al-ʿArabī, 2006.

ʿUthmānī, Shabbīr Aḥmad. *Mabādiʾ ʿilm al-ḥadīth wa-uṣūluhu*. Beirut: Dār al-Bashāʾir al-Islāmiyya, 2011.

al-Wāʾilī, Ḥasan b. Muḥammad. *Nuzhat al-albāb fī qawl al-Tirmidhī «wa-fī al-bāb»*. Jeddah: Dār Ibn al-Jawzī, 2005.

Yanofsky, Noson S. *The Outer Limits of Reason: What Science, Mathematics, and Logic Cannot Tell Us*. Cambridge, MA: The MIT Press, 2013.

Yazicioglu, Isra. "Redefining the Miraculous: al-Ghazālī, Ibn Rushd and Said Nursi on Qurʾanic Miracle Stories." *Journal of Qurʾanic Studies* 13, no. 2 (2011): 86–108.

al-Zabīdī, Murtaḍā. *Tāj al-ʿarūs min jawāhir al-Qāmūs*. Kuwait: Wizārat al-Irshād wa-l-Anbāʾ, 1965.

Zāhid, Muḥammad. *Ashraf al-tawḍīḥ: taqrīr urdū Mishkāt al-Maṣābīḥ*. Faisalabad: Maktabat al-ʿĀrifī, n.d.

Zakrewski, Sonia. "Life Expectancy." *UCLA Encyclopedia of Egyptology* 1, no. 1 (2015): 1–14.

Zaman, Iftikhar. "The Science of *Rijāl* as a Method in the Study of Hadiths." *Journal of Islamic Studies* 5, no. 1 (1994): 1–34.

Zaman, Muhammad Qasim. *The Ulama in Contemporary Islam: Custodians of Change*. Princeton: Princeton University Press, 2002.

Zaman, Muntasir. *Hadith Scholarship in the Indian Subcontinent: Aḥmad ʿAlī Sahāranpūrī and the Canonical Ḥadīth Literature*. Leicester: Qurtuba Books, 2021.

al-Zarkashī, Badr al-Dīn. *al-Laʾālī al-manthūra fī al-aḥādīth al-mashhūra*. Edited by Muḥammad Luṭfī al-Ṣabbāgh. Beirut: al-Maktab al-Islāmī, 1996.

Zaryūḥ, Muḥammad. *al-Muʿāraḍāt al-fikriyya al-muʿāṣira li-aḥādīth al-Ṣaḥīḥayn: dirāsa naqdiyya*. London: Takwīn li-l-Dirāsāt wa-l-Abḥath, 2020.

al-Zaylaʿī, ʿAbd Allāh b. Yūsuf. *Takhrīj aḥādīth al-Kashshāf*. Riyadh: Dār Ibn Khuzayma, 1414 AH.

al-Ziriklī, Khayr al-Dīn. *al-Aʿlām: qāmūs tarājim li-ashhar al-rijāl wa-l-nisāʾ min al-ʿArab wa-l-mustaʿribīn wa-l-mustashriqīn*. Beirut: Dār al-ʿIlm li-l-Malāyīn, 2002.

Zysow, Aron. *The Economy of Certainty: An Introduction to the Typology of Islamic Legal Theory*. Atlanta: Lockwood Press, 2013.

www.ingramcontent.com/pod-product-compliance
Lightning Source LLC
Chambersburg PA
CBHW031132090426
42738CB00008B/1065
* 9 7 8 1 9 1 5 0 2 5 3 2 6 *